New Aspects of
Quantity Surveying Practice

A text for all construction professionals

D1318934

Commissioning editor: Alex Hollingsworth
Development editor: Rebecca Rue
Production controller: Pauline Sones
Desk editor: Jackie Holding
Cover designer: Fred Rose

New Aspects of Quantity Surveying Practice

A text for all construction professionals

Duncan Cartlidge

The Scott Sutherland School,
The Robert Gordon University,
Aberdeen, UK

OXFORD   AUCKLAND   BOSTON   JOHANNESBURG
MELBOURNE   NEW DELHI

Butterworth-Heinemann
An imprint of Elsevier Science
Linacre House, Jordan Hill, Oxford OX2 8DP
225 Wildwood Avenue, Woburn MA 01801-2041

First published 2002

British Library Cataloguing in Publication Data
A catalogue record for this book is available from the British Library

Library of Congress Cataloguing in Publication Data
A catalog record for this book is available from the Library of Congress

ISBN 0 7506 5256X

Composition by Scribe Design, Gillingham, Kent
Printed and bound in Great Britain by MPG Books Ltd, Bodmin, Cornwall

Contents

List of figures and tables

Tables

Foreword:
Masters of the process

For which of you, intending to build a tower, sitteth not down first, and counteth the cost, whether he have sufficient to finish it? Lest haply, after he hath laid the foundation, and is not able to finish it, all that behold it begin to mock him, saying, this man began to build, and was not able to finish.

(St Luke 14 v. 28–30)

It is almost axiomatic that as long as people have wanted to build and have cared what things cost, there has always been a role for the quantity surveyor. At the root of the profession lies a detailed knowledge of the construction process – what things cost and more importantly, why they cost what they do. Based on this understanding, the quantity surveyor is in a position to advise clients on how to get best value for money from their capital expenditure.

There has been much talk in the industry that the confrontational nature of the industry is wasteful and that new approaches should be adopted to drive out this waste, including strategies such as partnering, alliance contracting, and cost-reimbursable target cost contracting. Frankly, this is just the beginning. Fundamental to any construction process is clients' understanding of their own business processes, being clear about what it is they want to build before fully engaging with the construction process.

Regardless of the procurement or delivery strategy employed, it is here that the quantity surveyor can add enormous value, helping to convert clients' needs into reality. The rest is just fashion.

As building design and construction techniques have become more and more complex, the design process has become more and more fragmented. As a result, clients find it an almost impossible industry with which to engage, facing a bewildering array of specialists. Architects, who used to dominate and were expected to lead the construction process, have largely lost control. Quantity surveyors on the other hand, whose detailed knowledge of construction used to be second to none – since it was they who had to understand, often interpret, measure and describe every detail of a structure as part of the construction process – have to an extent lost this skill as they have moved to fill the management vacuum left by the architects.

The industry, however, is about to change, not as result of fashionable initiatives, which are merely straws in the wind, but principally because of technological advances. Soon it will be possible to create complete project models to which all those involved in the process will have access, including clients. For example, simply by changing a construction detail, the whole life cost and the net present value of a structure will automatically change. There will be one integrated commercial management system around which the entire construction process will revolve; one system, linked to design and centred on powerful cost and time databases, will produce estimates and manage procurement, construction planning and administration, all linked to clients' own asset management databases. As a result of their training and intimate knowledge of construction process and cost, quantity surveyors, if they seize the opportunity, will inevitably become masters of the process.

Martin W. Bishop
Chairman, Franklin + Andrews
International Property and Construction Consultants

Preface

The Royal Institution of Chartered Surveyors' Quantity Surveying Think Tank, *Questioning the Future of the Profession*, heard
evidence that many within the construction industry thought
that chartered quantity surveyors were: arrogant, friendless and
unco-operative. In addition, they were perceived to add nothing
to the construction process, failed to offer services that clients
expected as standard, and too few had the courage to challenge
established thinking. In the same year, Sir John Egan called the
whole future of quantity surveying into question in the
Construction Industry Task Force report *Rethinking Construction* and, if this weren't enough, a report by the University of
Coventry entitled *Construction Supply Chain Skills Project*
concluded that quantity surveyors are 'arrogant and lacking in
interpersonal skills'. Little wonder then that the question was
asked, 'Will we soon be drying a tear over a grave marked "RIP
Quantity Surveying, 1792–2000"?' Certainly the changes that
have taken place in the construction industry during the past
twenty years would have tested the endurance of the most hardy
of beasts. Fortunately the quantity surveyor is a tough and
adaptable creature and, to quote and paraphrase Mark Twain,
'reports of the quantity surveyors' death are an exaggeration'.

I have spent the past 30 years or so as a quantity surveyor
in private practice, both in the UK and Europe, as well as
periods as a lecturer in higher education. During this time I
have witnessed a profession in a relentless search for an
identity, from quantity surveyor to building economist, to
construction economist, to construction cost advisor, to construction consultant, etc. I have also witnessed and been proud to be
a member of a profession that has always risen to a challenge
and has been capable of reinventing itself and leading from the

front, whenever the need arose. The first part of the twenty-first century holds many challenges for the UK construction industry as well as the quantity surveyor, but of all the professions concerned with the procurement of built assets, quantity surveying is the one that has the ability and skill to respond to these challenges.

This book is therefore dedicated to the process of transforming the popular perception that, in the cause of self-preservation, the quantity surveyor is wedded to a policy of advocating aggressive price-led tendering with all the problems that this brings, to one of a professional who can help deliver high-value capital projects on time and to budget with guaranteed life-cycle costs. In addition, it is hoped that this book will demonstrate beyond any doubt that the quantity surveyor is alive and well, adapting to the demands of construction clients and, what's more, looking forward to a long and productive future. Nevertheless, there is still a long hill to climb. During the production of this book I have heard major construction clients call the construction industry 'very unprofessional', and compare the role of the quantity surveyor to that of a 'post box'.

In an address to the Royal Institution of Chartered Surveyors in November 2001, the same Sir John Egan that had called the future of the quantity surveyor into question, but now as Chairman of the Egan Strategic Forum for Construction, suggested that the future for chartered surveyors in construction was to become process integrators, involving themselves in the process management of construction projects, and that those who clung to traditional working practices faced an uncertain future. The author would wholeheartedly agree with these sentiments:

The quantity surveyor is dead – long live the quantity surveyor – masters of the process!

Duncan Cartlidge
www.duncancartlidge.co.uk

Acknowledgements

My sincere thanks go to the following eminent individuals who have freely given their time and advice in the production of this book:

John Goodall of FIEC, Brussels.

Paul Braithwaite and Roger Simons of Cyril Sweett Ltd., London.

Ian Trushell of John Danskie Purdie & Partners, Glasgow.

Dr John Williams of TrustMarque for his contribution on e-commerce trust.

Tim Williams of Millstream Associates (Tenders Direct) for his contribution on e-procurement.

Bob Tragheim and Greg Watt of J Sainsbury plc. London, for their comments on supply chain management and partnering.

Dr Phil Durban of Medisense Inc., Abingdon, Oxon.

Alistair Gibb of Loughborough University.

Phil Hodson Information Systems, Director, Franklin + Andrews, London, for his contribution to Chapter Six on information systems in contract administion.

Crown copyright material is reproduced with the permission of the Controller of HMSO and Queen's Printer for Scotland.

Abbreviations

2D	Two-dimensional
3D	Three-dimensional
CAD	Computer aided design
CALS	Computer acquisition and lifetime support
CIC	Computer integrated construction
CPV	Common procurement vocabulary
CSF	Critical success factors
DXF	Data exchange format
EDI	Electronic data interchange
GPA	General procurement agreement
HTML	Hyper-text mark-up language
IAI	International Alliance for Interoperability
IFC	Industry foundation classes
IGES	Initial graphics exchange specification
IS	Information system
IT	Information technology
NAO	National Audit Office
OOP	Object-oriented programming
PDES	Product data exchange specification
PDM	Product data modelling
PFI	Private finance initiative
PPP	Public private partnerships
PPPP	Public private partnership programme
PSC	Public sector comparator
RICS	Royal Institution of Chartered Surveyors
SGML	Standard generalised mark-up language
SMEs	Small to medium sized enterprises
SPC	Special purpose company
SPV	Special purpose vehicle
SQL	Structured query language

STEP   Standards for the exchange of product data
SWOT   Strengths weaknesses opportunities threats analysis
VRML   Virtual reality mark-up language
XML   Extensible mark-up language

Dedication

This book is dedicated to the past: Chris Wilcox FRICS (1922–2000), perhaps the best measurer that I have known, or ever shall know; and to the future: Patrick and Dominic.

1

The catalyst of change

Introduction

This chapter examines the root causes of the changes that took place in the United Kingdom construction industry and quantity surveying practice during the latter half of the twentieth century. It sets the scene for the remaining chapters, which go on to describe how quantity surveyors can adapt to new markets and opportunities and contribute to client-led demands for added value.

The catalyst of change

The period between 1990 and 1995 will be remembered, as an eminent politician once remarked, as 'the mother of all recessions'. Certainly, from the perspective of the UK construction industry, this recessionary phase was the catalyst for many of the changes in working practices and attitudes that have been inherited by those who survived this period and continue to work in the industry. As chronicled in the following chapters, many of the pressures for change in the UK construction industry and its professions – including quantity surveying – have their origins in history, while others are the product of the rapid transformation in business practices that took place during the last two decades of the twentieth century. This book will therefore examine the background and causes of these changes, and then continue to analyse the consequences and effects on contemporary surveying practice.

Historical overview

The reasons underlying the remodelling of the UK construction industry during the first half of the 1990s are summarised in Figure 1.1. In the same way that the early nineteenth century saw the emergence of a number of factors that combined to bring about the Industrial Revolution and the 1950s appeared to be the moment in history to demolish the legacy of the industrial past and replace perfectly habitable low-rise housing with system-built tower blocks, 1990 was a watershed for the UK construction industry and its associated professions.

Figure 1.1 The heady brew of change.

As illustrated in Figure 1.1, by 1990 a 'heady brew of change' was being concocted on fires fuelled by the recession that was starting to have an impact on the UK construction industry. The main ingredients of this brew, in no particular order, were:

- The traditional UK hierarchical structure that manifested itself in a litigious, fragmented industry, where contractors and subcontractors were excluded from most of the design decisions.
- Changing patterns of workload due to the introduction of fee competition and compulsory competitive tendering.
- Widespread client dissatisfaction with the finished product.
- The emergence of privatisation and public private partnerships.
- The pervasive growth of information technology.
- The globalisation of markets and clients.

A fragmented and litigious industry

Boom and bust in the UK construction industry has been and will continue to be a fact of life for many years (see Figure 1.2), and much of the industry, including quantity surveyors, had learned to survive and prosper quite successfully in this climate.

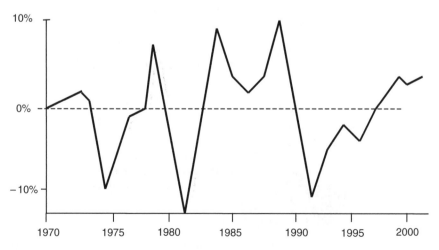

Figure 1.2 Construction output – percentage change 1970–2001 (source: Davis Langdon and Everest).

The rules were simple: in the good times a quantity surveyor earned fee income as set out in the Royal Institution of Chartered Surveyors' Scale of Fees for the preparation of, say, Bills of Quantities, and then in the lean times endless months or even years would be devoted to performing countless tedious re-measurements of the same work – once more for a fee. Contractors and subcontractors won work, albeit with very small profit margins, during the good times, and then when work was less plentiful they would turn their attentions to the business of the preparation of claims for extra payments for the inevitable delays and disruptions to the works. The standard forms of contract used by the industry, although heavily criticised by many, provided the impetus (if impetus were needed) to continue operating in this way. Everyone, including the majority of clients, appeared to be quite happy with the system, although in practice the UK construction industry was in many ways letting its clients down by producing buildings and other projects that were, in a high percentage of cases, over budget, over time and littered with defects. Time was running out on this system, and by 1990 the hands of the clock were at five minutes before midnight. A survey conducted in the mid 1990s by Property Week, a leading property magazine, among private sector clients who regularly commissioned new buildings or refurbished existing properties, provided a snapshot of the UK construction industry at that time. In response to the question 'Do projects finish on budget?', 30 per cent of those questioned replied that it was quite usual for projects to exceed the original budget (Figure 1.3).

In response to the question 'Do projects finish on time?', once more over 30 per cent of those questioned replied that it was

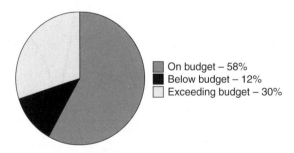

Figure 1.3 Survey of private sector construction clients' perceptions of budget accuracy.

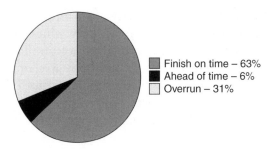

Finish on time – 63%
Ahead of time – 6%
Overrun – 31%

Figure 1.4 Survey of private sector construction clients' perception of completion times.

common for projects to overrun their planned completion by 1 or 2 months (Figure 1.4).

Parallels between the construction industry ethos at the time of this survey and the UK car industry of the 1960s make an interesting comparison. Austin, Morris, Jaguar, Rolls Royce, Lotus and marques such as the Mini and MG were all household names during the 1960s; today they are all either owned by foreign companies or out of business. Rover, now owned by Phoenix (UK), is the only remaining UK-owned carmaker, and itself was almost confined to history in 2000. Rover's decline from being the UK's largest carmaker in the 1960s is a living demonstration of how a country's auto industry can deteriorate, as well as being a stark lesson to the UK construction industry. The reasons behind the collapse of car manufacturing were flawed design, wrong market positioning, unreliability and poor build quality, lack of investment in new technology, and a failure to produce what the market demanded. Therefore, when the first Datsun cars began to arrive from Japan in the 1970s and were an immediate success, it was no surprise to anyone except the UK car industry. The British car buyer, after overcoming initial reservations about purchasing a foreign car, discovered a product that had nearly 100 per cent reliability, contained many features as standard that were extras on British-built cars, were delivered on time, and benefited from long warranties. Instead of producing what they perceived to be the requirements of the British car buyer, Datsun had researched and listened to the needs of the market, seen the failings of the home manufacturers, and then produced a car to meet them. Not only had the Japanese car industry researched the market fully; it had also invested in plant and machinery to increase build quality and reduce defects in their cars. Finally, the entire manufacturing process was analysed and a lean supply chain established to ensure the maximum economies of production. The scale of the improvements achieved in the car industry are impressive, with the time from completed design to launch reduced from 40 to 15 months, and the supplier defects to five parts per million.

So why by 1990 was the UK construction industry staring into the same abyss that the carmakers had faced 30 years earlier? In order to appreciate the situation that existed in the UK construction industry in the pre-1990 period it is necessary to examine the working practices of the UK construction industry, including the role of the contractor and the professions at this time. First, we will take a look back at recent history, and in particular at the events that took place in Europe in the first part of the nineteenth century and helped to shape UK practice (Goodall, 2000).

The UK construction industry – a brief history

Prior to the Napoleonic Wars, Britain, in common with its continental neighbours, had a construction industry based on separate trades. This system still exists in France as 'lots sépare', and variations of it can be found throughout Europe, including in Germany. The system works like this: instead of the multi-traded main contractor that operates in the UK, each trade is tendered for and subsequently engaged separately under the co-ordination of a project manager, or 'pilote'. In France smaller contractors usually specialise in one or two trades, and it is not uncommon to find a long list of contractors on the site board of a construction project.

The Napoleonic Wars, however, brought change, and nowhere more so than in Britain – the only large European state that Napoleon failed to cross or occupy. Paradoxically, the lasting effect the Napoleonic Wars had on the British construction industry was more profound than on any other national construction industry in Europe.

Whilst it is true that no military action actually took place on British soil, nonetheless the government of the day was obliged to construct barracks to house the huge garrisons of soldiers that were then being transported across the English Channel. As the need for the army barracks was so urgent and the time to prepare drawings, specifications, etc. was so short, the contracts were let on a 'settlement by fair valuation based on measurement after completion of the works'. This meant that constructors were given the opportunity and encouragement to innovate and to problem solve – something that was progressively withdrawn from them in the years to come. The same need for haste, coupled with the sheer magnitude of the individual projects, led to many contracts being let to a single builder or group of tradesmen 'contracting in gross', and the general contractor was born. When peace was made the Office of Works and Public Buildings, which had been increasingly concerned with the high cost of measurement and fair value procurement, in particular in the construction of Buckingham Palace and Windsor Castle, decided enough was enough. In 1828, separate trades contracting was discontinued for public works in England in favour of contracting in gross. The following years saw contracting in gross (general contracting) rise to dominate, and with this development the role of the builder as an innovator, problem solver and design team member was stifled to the point where contractors operating in the UK system were reduced to simple executors of the works and instructions (although in Scotland the separate trades system survived until the early 1970s). However, history had another twist, for in 1834 architects decided that they wished to divorce themselves from surveyors and establish the Royal Institute of British Architects (RIBA), exclusively for architects. The grounds for this great schism were that architects wished to distance themselves from surveyors and their perceived obnoxious commercial interest in construction. The top-down system that characterises so much of British society was stamped on the construction industry. As with the death of separate trades contracting, the establishment of the RIBA ensured that the UK contractor was once again discouraged from using innovation. The events of 1834 were also responsible for the birth of another UK phenomenon, the quantity surveyor, and for another unique feature of the UK construction industry – post construction liability.

The ability of a contractor to re-engineer a scheme design in order to produce maximum buildability is a great competitive advantage, particularly on the international scene (see Table 1.1). As discussed in Chapter 2, a system of project insurance that is already widely available on the Continent is starting to make an appearance in the UK; with this, the design and execution teams can safely circumvent their professional indemnity insurance and operate as partners under the protective umbrella of a single policy of insurance, thereby allowing the interface of designers and contractors.

However, back to history. For the next 150 or so years the UK construction industry continued to develop along the lines outlined above, and consequently by the third quarter of the twentieth century the industry was characterised by powerful professions carrying out work on comparatively generous fee scales, contractors devoid of the capability to analyse and refine design solutions, forms of contract that made the industry one of the most litigious in Europe, and procurement systems based upon competition and selection by lowest price and not value for money. Some within the industry had serious concerns about procurement routes and documentation, the forms of contract in use leading to excess costs, suboptimal building quality and time delays, and the adversarial and conflict-ridden relationships between the various parties. A series of government-sponsored reports (Simon, 1944; Emmerson, 1962; Banwell, 1964) attempted to stimulate debate about construction industry practice, but with little effect.

It was not just the UK construction industry that was obsessed with navel-gazing during the last quarter of the twentieth century; quantity surveyors had also been busy penning numerous reports into the future prospects for their profession. The most notable of which were: The Future Role of the Chartered Quantity Surveyor (1983), Quantity Surveying 2000 – The Future Role of the Chartered Quantity Surveyor (1991) and the Challenge for Change: QS Think Tank (1998), all produced either directly by, or on behalf of, The Royal Institution of Chartered Surveyors.

The 1971 report, The Future Role of the Quantity Surveyor (RICS), was the product of a questionnaire sent to all firms in private practice together with a limited number of public sector organisations; sadly, but typically, the survey resulted in a mere 35% response rate. The report paints a picture of a world where the quantity surveyor was primarily a producer of Bills of Quantities; indeed, the report comes to the conclusion that the distinct competence of the quantity surveyor of the 1970s was measurement – a view, it should be added, still shared by many today. In addition, competitive single stage tendering was the norm, as was the practice of receiving most work via the patronage of an architect. It was a profession where design and construct projects were rare, and quantity surveyors were discouraged from forming multidisciplinary practices and encouraged to adhere to the scale of fees charges. The report observes that clients were becoming more informed, but there was little advice about how quantity surveyors were to meet this challenge. A mere 25 years later the 1998 report, The Challenge of Change, was drafted in a business climate driven by information technology, where quantities generation is a low-cost activity and the client base is demanding that surveyors demonstrate added value. In particular, medium-sized quantity surveying firms (i.e. between 10 and 250 employees) were singled out by this latest report to be under particular pressure owing to difficulties in:

* Competing with the large practices' multiple disciplines and greater specialist knowledge
* Attracting and retaining a high quality work force
* Achieving a return on the necessary investment in IT
* Competing with the small firms with low overheads.

Consequently, the surveying profession has been predicted to polarise into two groups; the large multidisciplinary practices capable of matching the problem solving capabilities of the large accountancy-based consulting firms, and small practices that can offer a fast response from a low cost base for clients, as well as providing services to their big brother practices. Interestingly, The Challenge for Change report also predicts that the distinction between contracting and professional service organisations will blur, perhaps with a move towards the methods of the French construction industry, where contractors are the largest employers of construction professionals – a quantum leap from the 1960s, when chartered surveyors were forced to resign from their institution if they worked for contracting organisations!

The British system compared

An interesting and unique opportunity directly to compare the British system of procuring and managing projects with that of a European neighbour was provided during the construction of the Channel Tunnel. In the early 1990s, Eurotunnel needed a new security scanning system to protect the fixed link – the Euroscan facility. It commissioned a leading UK architectural practice for design on both sides of the channel, and procured medium-sized

British and French firms for the construction. Research under-taken by Graham Winch and Andrew Edkins of Bartlett School of Graduate Studies, University College London, with the enthu-siastic co-operation of Eurotunnel plc and funding by the Engineering and Physical Sciences Research Council, gave a unique opportunity to compare project performance in the two countries with a functionally equivalent building, a common design and a single client. Table 1.1 shows the performance on the two sides of the channel, and how the French performed much better than the British. Both project teams faced the similar challenges, largely generated by problems with scanning technol-ogy, yet the French team coped with them more smoothly. Why?

The answer would seem to lie in the differences in the organ-isation of the two projects:

- The French contract included detail design, the norm in France; the British contractor was deemed not capable of entering into a design and build contract due to the require-ments for design information under the JCT 80.
- The French contractor re-engineered the project, simplifying the design and taking out unnecessary costs. This was possi-ble because of the single point project liability that operates in France.

Table 1.1 British and French construction performance compared

Performance indicator	French performance	British performance
Design costs	£323 525	£465 000
Contractor tender price	£3 852 754	£3 897 000
Contractor out-turn cost	£4 178 652	£4 482 375
Total acquisition cost	£4 502 178	£4 947 375
Contractor cost increase	8.5%	15%
Contract programme	Equal	Equal
Programme overrun	0%	28%
Site management staff	4	8
Procurement	Lump sum – bespoke contract	Approximate Bill of Quantities – JCT 80
Strengths/weaknesses	Contractor's engineering capacity means value engineering as standard practice	Liability insurance prevents value engineering; process complexity

Source: Bartlett School of Graduate Studies, University College London.

- Under the French contract, the British architect could not object to these contractor led changes. Under JCT 80, professional indemnity considerations meant that the architect refused to allow the British contractor to copy the French changes.
- The simplified French design was easier, cheaper and quicker to build. This meant that there was room for manoeuvre as the client induced variations mounted, whereas the British run project could only cope by increasing programme and budget. Once the project began to run late, work on construction became even less effective as the team had to start working out of sequence around the installation of scanning equipment.

The researcher's conclusion is that British procurement arrangements tend to generate complexity in project organisation, while the engineering capabilities of French contractors mean that they are able to simplify the design. Indeed, they argue that it is these capabilities that are essential to the French contractor's ability to win contracts.

For many observers the question of single point or project liability – the norm in many countries, such as Belgium and France – is pivotal in the search for adding value to the UK construction product, and is at the heart of the other construction industries' abilities to re-engineer designs. Single point project liability insurance is insurance that protects all the parties involved in both the design and the construction process against failures in both design and construction of the works for the duration of the policy. The present system, where some team members are insured and some not, results in a tendency to design defensively, caveat all statements and advice with exclusions of liability, and seek help from no other members of the team – not a recipe for team work. In the case of a construction management contract, the present approach to latent defect liability can result in the issue of 20–30 collateral warranties, which facilitates the creation of a contractual relationship where one would otherwise not exist in order that the wronged party is then able to sue under contract rather than rely on the tort of negligence. Therefore in order to give contractors the power truly to innovate and to use techniques like value engineering (see Chapter 2), there has to be a fundamental change in the approach to liability. Contract forms could be amended to allow the contractor to modify the technical design prior to construction, with the consulting architects and engineers waiving their rights to interfere.

If this approach is an option, then why does the UK construction industry still fail to produce the goods? The new Wembley Stadium and Pickett's Lock, both proposed venues for the World Athletics Championship in 2005, are prime examples of the traditional UK approach – namely, this is the design, this is the cost, that's it, we've had it! The principal problems behind the failure of these two high-profile fiascos were no business case, little or no understanding of the needs of the client, and the inability of a contractor to re-engineer the proposals and produce alternatives. The result – grandiose designs with large price tags and a complete disregard of the need to pay back the cost of the project from revenues generated by the built asset, in this case a sports stadium.

Opponents of the proposal to introduce single-point liability cite additional costs as a negative factor. However, indicative costs given by Royal & SunAlliance seem to prove that these are minimal – for example, traditional structural and weatherproofing: 0.65–1.00% of contract value total cover, including structural, weatherproofing, non-structural and M&E; 1–2% of contract value to cover latent defects for periods of up to 12 years, to tie in with the limitations provisions of contracts under Seal. As in the French system, technical auditors can be appointed to minimise risk and, some may argue, add value through an independent overview of the project.

Changing patterns of workload

The patterns of workload that quantity surveyors had become familiar with were also due to change. The change came chiefly from two sources:

1. Fee competition and compulsory competitive tendering (CCT)
2. The emergence of a new type of construction client.

Fee competition and compulsory competitive tendering

Until the early 1970s, fee competition between professional practices was almost unheard of. All the professional bodies published scales of fees, and competition was vigorously discouraged on the basis that a client engaging an architect, engineer or surveyor should base his or her judgement on the type of service and not on the level of fees. Consequently, all professionals within a specific discipline quoted the same fee. However, things were to change with the election of the Conservative Government in 1979. The new government introduced fee competition into the public sector by way of its compulsory competitive tendering programme, and for the first time professional practices had to compete for work in the same manner as contractors or subcontractors – i.e. they would be selected by competition, mainly on the basis of price. The usual procedure was to submit a bid based upon scale of fees minus a percentage. Initially these percentage reductions were a token 5 or 10%, but as work became difficult to find in the early 1980s, practices offered 30 or even 40% reduction on fee scales. It has been suggested that during the 1980s fee income from some of the more traditional quantity surveying services was cut by 60%. Once introduced there was no going back, and soon the private sector began to demand the same reduction in fee scales; within a few years the cosy status quo that had existed and enabled private practices to prosper had gone. The Monopolies and Mergers Commission's 1977 report into scales of fees for surveyors' services led the Royal Institution of Chartered Surveyors to revise its byelaws in 1983 to reduce the influence of fee scales to the level of 'providing guidance' – the gravy train had hit the buffers!

Byelaw 24 was altered from:

No member shall with the object of securing instructions or supplanting another member of the surveying profession, knowingly attempt to compete on the basis of fees and commissions

to

... no member shall ... quote a fee for professional services without having received information to enable the member to assess the nature and scope of the services required.

A review of more than 250 quantity surveying practices, carried out in 1999 by Mirza & Nacey Research for Building Magazine, showed that the average fee for quantity surveyors (expressed as a percentage of construction cost) over a range of new build projects was just 1.7%! As a result, professional practices found it increasingly difficult to offer the same range of services and manning levels on such a reduced fee income; they had radically to alter the way they operated, or go out of business. However, help was at hand for the hard-pressed practitioner; the difficulties of trying to manage a practice on reduced fee scale income during the later part of the 1980s were mitigated by a property boom, which was triggered in part by a series of government-engineered events that combined to unleash a feeding frenzy of property development. In 1988 construction orders peaked at £26.3 billion, and the flames under the heady brew of change were dampened down, albeit only for a few years. The most notable of these events were:

- The so-called Stock Exchange 'Big Bang' of 1986, which had the direct effect of stimulating the demand for high-tech offices
- The deregulation of money markets in the early 1980s, which allowed UK banks for the first time to transfer money freely out of the country, and foreign finance houses and banks to lend freely on the UK market and invest in UK real estate
- The announcement by the Chancellor of the Exchequer, Nigel Lawson, of the abolition of double tax mortgage relief for domestic dwellings in 1987, which triggered an unprecedented demand for residential accommodation; the result was a massive increase in lending to finance this sector, as well as spiralling prices and land values
- Last but by no means least, the relaxation of planning controls, which left the way open for the development of out-of-town shopping centres and business parks.

However, most property development requires credit, and the boom in development during the late 1980s could not have taken place without financial backing. By the time the hard landing came in 1990, many high street banks with a reputation for prudence found themselves dangerously exposed to high-risk real estate projects. During the late 1980s, virtually overnight the banks changed from conservative risk managers to target-driven loan sellers, and by 1990 they found themselves with a total property related debt of £500 billion. The phenomenon was not just confined to the UK. In France, for example, one bank alone, Credit Lyonnais, was left with ff300 billion of unsecured loss after property deals on which the bank had lent money collapsed because of over supply and a lack in demand; only a piece of creative accountancy and state intervention saved the French bank from insolvency. The property market crash in the early 1990s occurred mainly because investors suffered a lack of confidence in the ability of real estate to provide a good return on investment in the short to medium term in the light of high interest rates, even higher mortgage rates, and an inflation rate that doubled within 2 years. In part it was also brought about by greed because of the knowledge that property values had historically seldom delivered negative values.

The emergence of a new type of construction client

Another vital ingredient in the brew of change was the emergence of a new type of construction client. Building and civil engineering works have traditionally been commissioned by either public or private sector clients. The public sector has been a large and important client for the UK construction industry and its professions. Most government bodies and public authorities would compile lists or 'panels' of approved quantity surveyors and contractors for the construction of hospitals, roads and bridges, social housing, etc., and inclusion on these panels ensured that they received a constant and reliable stream of work. However, during the 1980s the divide between public and private sectors was to blur. The Conservative government of 1979 embarked upon an energetic and extensive campaign of the privatisation of the public sector that culminated in the introduction of the Private Finance Initiative in 1992 (see Chapter 4). Within a comparatively short period there was a shift from a system dominated by the public sector to one where the private sector was growing in importance. Despite this shift to the private sector the public sector still remains influential; in 1999, for example, it accounted for 37% or £7.5 billion of the UK civil engineering and construction industry's business, with a government pledge to maintain this level of expenditure. Nevertheless, the privatisation of the traditional public sector resulted in the emergence of major private sector clients such as the British Airports Authority, privatised in 1987, with an appetite for change and innovation. This new breed of client was, as the RICS had predicted in its 1971 report on the future of quantity surveying, becoming more knowledgeable about the construction process, and such clients were not prepared to sit on their hands while the UK construction industry continued to under perform. Clients such as Sir John Egan, who in July 2001 was appointed Chairman of the Strategic Forum of Construction, became major players in the drive for value for money. The poor performance of the construction industry in the private sector has already been examined (Figures 1.3 and 1.4). However, if anything, performance in the public sector paints an even more depressing picture. This performance was scrutinised by the National Audit Office (NAO) in 2001 in its report Modernising Construction (Auditor General, 2000), which found that the vast majority of projects were over budget and delivered late (Figures 1.5, 1.6). So dire has been the experience of some public sector clients – for example the Ministry of Defence – that radical new client driven initiatives for procurement, such as prime contracting (see Chapter 2) have been introduced.

In particular there were a number of high-profile public projects, like the disastrous British Library delivered over 10

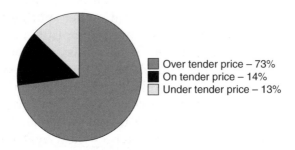

Figure 1.5 Public sector construction budget performance.

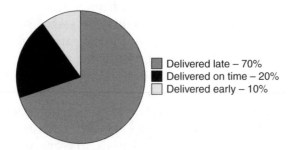

Figure 1.6 Public sector construction delivery performance.

years late, and the Faslane Trident Submarine Berth which increased in cost by more than 200% and was delivered 2½ years late. If supply chain communications were polarised and fragmented in the private sector, then those in the public sector were even more so. A series of high-profile cases in the 1970s, in which influential public officials were found to have been guilty of awarding construction contracts to a favoured few in return for bribes, instilled paranoia in the public sector, which led to it distancing itself from contractors, subcontractors and suppliers – in effect from the whole supply chain. At the extreme end of the spectrum this manifested itself in public sector professionals refusing to accept even a diary, calendar or a modest drink from a contractor in case it was interpreted as an inducement to show bias. In the cause of appearing to be fair, impartial and prudent with public funds, most public contracts were awarded as a result of competition between a long list of contractors on the basis of the lowest price. The 2001 National Audit Office report suggests that the emphasis on selecting the lowest price is a significant contributory factor to the tendency towards adversarial relationships. Attempting to win contracts under the 'lowest price wins' mentality leads firms to price work unrealistically low and then seek to recoup their profit margins through contract variations arising from, for example design changes and other claims leading to disputes and litigation.

In an attempt to eradicate inefficiencies the public sector commissioned a number of studies such as The Levene Efficiency Scrutiny in 1995, which recommended that departments in the public sector should:

- Communicate better with contractors to reduce conflict and disputes
- Increase the training that their staff receive in procurement and risk management
- Establish a single point for the construction industry to resolve problems common to a number of departments. The lack of such a management tool was identified as one of the primary contributors to problems with the British Library project.

In June 1997 it was announced that Compulsory Competitive Tendering would be replaced with a system of Best Value in order to introduce, in the words of the local government minister

Hilary Armstrong, 'an efficient, imaginative and realistic system of public sector procurement'. Legislation was passed in 1999, and from the 1 April 2000 it became the statutory duty of the public sector to obtain best value. Best value will be discussed in more detail in Chapter 4.

The information technology revolution

As measurers and information managers, quantity surveyors have been greatly affected by the information technology revolution. Substantial parts of the chapters that follow are devoted to the influence that IT has had and will continue to have, both directly and indirectly, on the quantity surveying profession. However, this opening chapter would not be complete without a brief mention of the contribution of IT to the heady brew of change. To date, mainly individual IT packages have been used or adapted for use by the quantity surveyor – for example, spreadsheets. However, the next few years will see the development of IT packages designed specifically for tasks such as measurement and quantification, which will fundamentally change working practices. The speed of development has been breathtaking. In 1981 the Department of the Environment developed and used a computer-aided bill of quantities production package called 'Enviro'. This then state-of-the-art system required the quantity surveyor to code each measured item, and on completion the codes were sent to Hastings, on the South Coast of England, where a team of operators would input the codes, with varying degrees of accuracy, into a mainframe computer. After the return of the draft bill of quantities to the measurer for checking, the final document was then printed, which in most cases was 4 weeks after the last dimensions were taken off!

In recent years architects have made increasing use of computer-aided design (CAD) in the form of 2D drafting and 3D modelling for the production of project information. A report published in 2000 by the Construction Industry Computing Association and entitled Architectural IT Usage and Training Requirements indicated that in practices with more than six staff, between 95 and 100% of all those questioned used 2D drafting to produce information. This shift from hand drawn drafting to IT-based systems has allowed packages to be developed that link the production of drawings and other information to their measurement and quantification, thereby revolutionising the once labour-intensive bill of quantities preparation procedure. Added to this, the spread of the digital economy means that drawings and other project information can be produced, modified and transferred globally. One of the principal reasons for quantity surveyors' emergence as independent professionals during the Napoleonic Wars and their subsequent growth to hold a pivotal role in the construction process had, by the end of the 1990s, been reduced to a low cost IT operation.

Those who mourn the demise of traditional methods of bill of quantities production should at least take heart that no longer will the senior partner be able to include those immortal lines in a speech at the annual Christmas office party – 'you know after twenty years of marriage my wife thinks that quantity surveying is all about taking off and working up' – pause for laughter!

As mentioned previously, there had been serious concern both in the industry and in government about the public image of UK Construction plc. The 1990 recession had opened the wounds in the construction industry and shown its vulnerability to market pressures. Between 1990 and 1992 over 3800 construction enterprises became insolvent, taking with them skills that would be badly needed in the future. The professions also suffered a similar haemorrhage of skills as the value of construction output fell by double digit figures year on year. The recession merely highlighted what had been apparent for years: the UK construction industry and its professional advisors had to change. The heady brew of change was now complete, but concerns over whether or not the patient realised the seriousness of the situation still gave grounds for concern. The message was clear; industry and quantity surveying must change or, like the dinosaur, be confined to history!

Response to change

In traditional manner, the UK construction industry turned to a report to try to solve its problems. In 1993 Sir Michael Latham, an academic and politician, was tasked to prepare yet another review, this time of the procurement and contractual arrangements in the United Kingdom construction industry. In July 1994, Constructing The Team (or The Latham Report, as it became known) was published. The aims of the initiative were to reduce conflict and litigation, as well as to improve the industry's productivity and competitiveness. The construction industry held its breath – was this just another Banwell or Simon to be confined, after a respectful period, to gather dust on the shelf? Thankfully not! The UK construction industry was at the time of publication in such a fragile state that the report could not be ignored. That's not to say that it was greeted with open arms by everyone – indeed, the preliminary report Trust and Money, produced in December 1993, provoked profound disagreement in the industry and allied professions.

Latham's report found that the industry required a good dose of medicine, which the author contended should be taken in its entirety if there was to be any hope of a revival in its fortunes. The Latham Report highlighted the following areas as requiring particular attention to assist UK construction industries to become and be seen as internationally competitive:

- Better performance and productivity, to be achieved by using adjudication as the normal method of dispute resolution, the adoption of a modern contract, better training, better tender evaluation, and the revision of post-construction liabilities to be more in line with, say, France or Spain, where all parties and not just the architect are considered to be competent players and all of them therefore are liable for non-performance for up to 10 years
- The establishment of well-managed and efficient supply chains and partnering agreements
- Standardisation of design and components, and the integration of design, fabrication and assembly to achieve better buildability and functionality
- The development of transparent systems to measure performance and productivity both within an organisation and with competitors
- Teamwork and a belief that every member of the construction team from client to subcontractors should work together to produce a product of which everyone can be justifiably proud.

The Latham Report placed much of the responsibility for change on clients in both public and private sectors. For the construction industry, Latham set the target of a 30% real cost reduction by year 2000, a figure based on the CRINE (Cost Reduction Initiative for the New Era) review carried out in the oil and gas industries a few years previously (CRINE, 1994). The CRINE review was instigated in 1992, with the direct purpose of identifying methods by which to reduce the high costs in the North Sea oil and gas industry. It involved a group of operators and contractors working together to investigate the cause for such high costs in the industry, and also to produce recommendations to aid the remedy of such. The leading aim of the initiative was to reduce development and production costs by 30%, this being achieved through recommendations such as the use of standard equipment, simplifying and clarifying contract language, removing adversarial clauses, rationalisation of regulations, and the improvement of credibility and quality qualifications. It was recommended that the operators and contractors work more closely, pooling information and knowledge, to help drive down the increasing costs of hydrocarbon products and thus indirectly promote partnering and alliancing procurement strategies (see Chapter 3). The CRINE initiative recommendations were accepted by the oil and gas industries, and it is now widely accepted that without the use of partnering/alliancing a great number of new developments in the North Sea would not have been possible. Shell UK Exploration and Production reported that the performance of the partners in the North Fields Unit during the period 1991–1995 resulted in an increase in productivity of 25%, a reduction in overall maintenance costs of 31% in real terms, and a reduction in platform 'down time' of 24%. Could these dramatic statistics be replicated in the construction industry? 'C' is not only for construction but also for conservative, and many sectors of the construction industry considered 30% to be an unrealistically high and unreachable target. Nevertheless, certain influential sections of the industry, including Sir John Egan and BAA, accepted the challenge and went further declaring that 50 or even 60% savings were achievable. It was the start of the client-led crusade for value for money.

The Latham Report spawned a number of task groups to investigate further the points raised in the main report, and in October 1997, as a direct result of one of these groups, Sir John Egan, a keen advocate of Sir Michael Latham's report and known to be a person convinced of the need for change within the industry, was appointed as head of the Construction Task Force. One of the Task Force's first actions was to visit the Nissan UK car plant in Sunderland to study the company's supply chain management techniques and to determine whether they could be utilised in construction (see Chapter 2). In June 1998 the Task Force published the report Rethinking Construction (DoE, 1998), which was seen as the blueprint for the modernisation of the systems used in the UK construction industry to procure work. As a starting point, Rethinking Construction revealed that in a survey of major UK property clients, many continued to be dissatisfied with both contractors' and consultants' performance. Added to the now familiar concerns about failure to keep within agreed budgets and completion schedules, clients revealed that:

- More than a third of them thought that consultants were lacking in providing a speedy and reliable service
- They felt they were not receiving good value for money insofar as construction projects did not meet their functional needs and had high whole-life costs
- They felt that design and construction should be integrated in order to deliver added value.

As for the quantity surveyors, the 1990s ended with perhaps the unkindness cut of all. The RICS, in its Agenda for Change initiative, replaced its traditional divisions (which included the Quantity Surveying Division) with 16 faculties, not unlike the system operated by Organisme Professionel de Qualification Technique des Economistes et Coordonnateurs de la Construction (OPQTECC), the body responsible for the regulation of the equivalent of the quantity surveyor in France. It seemed to some that the absence of a quantity surveying faculty would result in the maginalisation of the profession; however, the plan was implemented in 2000, with the Construction Faculty being identified as perhaps the new home for the quantity surveyor within the RICS.

Beyond the rhetoric

How are the construction industry and the quantity surveyor rising to the challenges outlined in the previous pages?

When the much-respected quantity surveyors Arthur J. and Christopher J. Willis penned the foreword to the eighth edition of their famous book Practice and Procedure for the Quantity

Surveyor in 1979, the world was a far less complicated place. Diversification into new fields for quantity surveyors included heavy engineering, coal mining and 'working abroad'. In the Willis's book, the world of the quantity surveyor was portrayed as a mainly technical back office operation providing a limited range of services where, in the days before compulsory competitive tendering and fee competition, 'professional services were not sold like cans of beans in a supermarket'. The world of the Willis's was typically organised around the production of bills of quantities and final accounts, with professional offices being divided into pre- and post-contract services. This model was uniformly distributed across small and large practices, the main difference being that the larger practices would tend to get the larger contracts and the smaller practices the smaller contracts. This state of affairs had its advantages, as most qualified quantity surveyors could walk into practically any office and start work immediately; the main distinguishing feature between practices A and B was usually only slight differences in the format of taking-off paper. However, owing to the changes that have taken place not only within the profession and the construction industry but also on the larger world stage (some of which have been outlined in this chapter), the world of the Willis's has, like the British Motor Car Industry, all but disappeared forever.

In the early part of the twenty-first century, the range of activities and sectors where the quantity surveyor is active is becoming more and more diverse. The small practice concentrating on traditional pre- and post-contract services is still alive and healthy. However, at the other end of the spectrum the larger practices are now rebadged as international consulting organisations and would be unrecognisable to the Willis's. The principal differences between these organisations and traditional large quantity surveying practices are generally accepted to be the elevation of client focus and business understanding and the move by quantity surveyors to develop clients' business strategies and deliver added value. As discussed in the following chapters, modern quantity surveying involves working in increasingly specialised and sectorial markets where skills are being developed in areas including strategic advice in the PFI, partnering, value and supply chain management.

From a client's perspective it is not enough to claim that the quantity surveyor and/or the construction industry is delivering a better value service; this has to be demonstrated. Certainly there seems to be a move by the larger contractors away from the traditional low-profit, high-risk, confrontational procurement paths toward deals based on partnering and PFI and the team approach advocated by Latham. A survey in November 2000 carried out by Hewes Associates for Building Magazine clearly illustrates this shift (see Table 1.2).

The terms of reference for the Construction Industry Task Force concentrated on the need to improve construction efficiency and to establish best practice. The industry was urged to take a lead from other industries, such as car manufacturing, steel making, food retailing and offshore engineering, as examples of market sectors that had embraced the challenges of rising world-class standards and invested in and implemented lean production techniques. Rethinking Construction identified five driving forces that needed to be in place to secure improvement in construction and four processes that had to be significantly enhanced, and set seven quantified improvement targets, including annual reductions in

Table 1.2 Trends in procurement

Contractors' workload

| Company | Turnover: | | | | | |
| | Total turnover (£ million) | | Traditional procurement (%) | | Partnering, PFI, etc. (%) | |
	1998	1999	1998	1999	1998	1999
John Mowlem	810.6	983.6	48.4	30.6	33.6	56.2
Morgan Sindall	387.3	409.3	95.6	77.7	0.0	20.4
Ballast Wiltshire	301.7	403.0	76.4	61.3	23.6	38.7
Sheppard	238.3	269.4	88.6	54.2	11.4	45.8
Norwest Holst	200.4	177.8	82.9	56.1	5.4	30.7
Wates Construction	211.0	312.4	97.1	87.3	2.9	12.7
M.J. Gleeson	219.1	280.0	44.0	32.1	49.9	61.5
Willmott Dixon	284.0	277.0	68.0	54.9	23.6	35.4
Galliford	180.0	272.0	37.8	26.1	44.0	54.8
John Sisk & Son	127.5	131.0	58.4	42.1	41.6	57.9
Carillion	1470.0	1477.0	48.0	42.9	41.2	45.6
Kvaerner/Skanska	628.9	745.5	72.0	68.0	25.3	29.6
Mansell	417.9	442.8	72.3	69.5	16.9	19.3

Source: Building Magazine

Table 1.3 Rethinking construction recommendations

The five key drivers that need to be in place to achieve better construction are:

1. Committed leadership
2. Focus on the customer
3. Integration of process and team around the project
4. A quality-driven agenda
5. Commitment to people.

The four key projected processes needed to achieve change are:

1. Partnering the supply chain – development of long-term relationships based on continuous improvement with a supply chain
2. Components and parts – a sustained programme of improvement for the production and the delivery of components
3. Focus on the end product – integration and focusing of the construction process on meeting the needs of the end user
4. Construction process – the elimination of waste.

The seven annual targets capable of being achieved in improving the performance of construction projects are:

1. To reduce capital costs by 10%
2. To reduce construction time by 10%
3. To reduce defects by 20%
4. To reduce accidents by 20%
5. To increase the predictability of projected cost and time estimates by 10%
6. To increase productivity by 10%
7. To increase turnover and profits by 10%.

construction costs and delivery times of 10% and reductions in building defects of 20%.

The report also drew attention to the lack of firm quantitative information with which to evaluate the success or otherwise of construction projects. Such information is essential for two purposes:

1. To demonstrate whether completed projects have achieved the planned improvements in performance
2. To set reliable targets and estimates for future projects based on past performances.

It has been argued that organisations like the Building Cost Information Service have been providing a benchmarking service for many years through its tender based index. Additionally, what is now required is a transparent mechanism to enable clients to determine for themselves which professional practice, contractor, subcontractor, etc. delivers best value.

In an attempt to remedy this information gap, key performance indicators (KPIs) were introduced in 1999 in order to raise awareness of measuring performance. Although criticised in some quarters as simplistic, the government departments responsible for the scheme recognise that KPIs are not a substitute for more comprehensive benchmarking assessments. They do, however, enable organisations to gauge their performance in relation to other companies. KPIs start from the point that clients want their projects delivered – on time, on budget, free from defects, efficiently, right first time, safely and by profitable companies. The KPI framework consists of seven main groups: time, cost, quality, client satisfaction, client changes, business performance, and health and safety. For example, if the KPIs for cost were applied to measure cost performance of a typical project, the following results are a sample of what may be obtained. Over the range of indicators, the references are to the key project stages shown in Figure 1.7.

Using this framework enables the measurement of project and organisational performance, and the information gained can also be used for benchmarking purposes as a move towards determining best practice. For example, construction clients will be able to assess the suitability of an organisation to be included in a design team by asking it to provide information about how it performs against a range of indicators. Similar benchmarking exercises carried out in Europe – for example, The European Construction Benchmarking Pilot Study, concluded in June 2001 by the FIEC and the Building Research Establishment – have met with little enthusiasm from European construction organisations due, it was thought, to the lack of client demand in what is perceived to be an efficient industry, certainly when compared to the UK.

Measuring the performance of consultants

In June 2001, the Construction Best Practice Programme introduced performance indicators for construction consultants. The

Figure 1.7 Key performance indicators – cost

Indicator	Definition	Outcome
1. Cost of construction	Change in the construction cost in cost/m^2 at point B compared with 1 year earlier	–2.1%
2. Cost predictability – design	Change between actual design cost at point C and the estimated cost at point A	–2.0%
3. Cost predictability – construction	Change between the actual construction cost at point C and the estimated construction cost at point A	+2.2%
4. Cost predictability – design and construction	Change between the actual cost at point C and the estimated cost at point A	–0.9%
5. Cost predictability – construction: client variations	Changes attributable to client variations	+1.8%
6. Cost predictability – construction: design team variations	Changes attributable to design team variations	+0.4%
7. Cost to rectify defects in maintenance period	Cost of making good defects expressed as a percentage of cost at point C	+4.3%
8. Whole life costs	The annual operating and maintenance cost following point C expressed as a percentage of the actual design and construction cost at point C	+2.5%

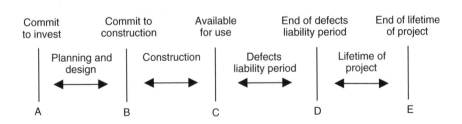

Source: DETR

government sponsored project partners include the Royal Institution of Chartered Surveyors and the Royal Institute of British
Architects, and the project aims to provide performance indicators for:

- Client satisfaction with the services provided by a firm of
  consultants, which include clients' perceptions of: overall
  performance, timely delivery, value for money, quality, and
  health and safety awareness. This form of performance
  assessment is of a narrow focus, whereas a wider assessment
  can be provided by:
- Company performance, which includes how the company
  compares to the sector as a whole in terms of: training,
  productivity and profitability.

In the first case, information can be obtained directly from
clients by, for example, asking how satisfied a client was that
the consultant provided a quality service on completed commission and using a 1–10 scale where 1 = totally dissatisfied and 10
= totally satisfied. In the second case, each year information that
is submitted by construction consultancies is gathered from
Companies House by the DETR. From this information the KPIs
are calculated over a range of services and performances as
indicated above and illustrated graphically. Consultants' KPIs
can be used in the following ways:

- Clients can use KPIs to promote continuous performance in
  partnership with consultants
- At the outset of a new project, KPIs can be used to set performance targets
- Throughout a project KPIs can be used to monitor progress
- As a marketing tool, to demonstrate the commitment to
  continuous improvement – for so long missing in the
  construction process, but taken as the norm in other sectors.

The usual method used to give an overall picture of a company's
performance is a radar chart.

A similar benchmarking exercise has been carried out by the
FIEC and the BRE, focusing on the European Construction
Industry. Figure 1.8 is a radar chart that illustrates the plot of
a month's performance on a French construction site compared
with the European average.

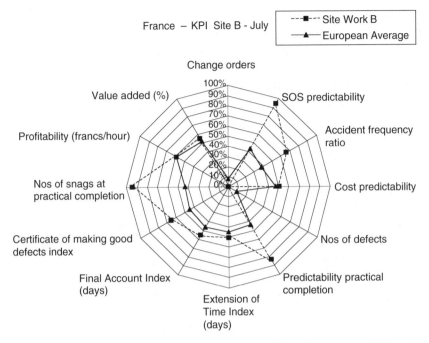

Figure 1.8 Radar chart comparing 1-month performance of French construction site with the European average (source: FIEC).

From the chart it can be determined that:

• The number of defects (snags) at practical completion is more than twice as many as the EU average
• Accident frequency ratio is higher than the EU average
• Remaining KPIs are better than average
• Start on-site predictability and predicted practical completion are the best performing KPIs
• Profitability and value added are comparable with the EU average.

Finally, to add to the call for leaner/added value in construction, the National Audit Office analysis of the most influential studies carried out during the 1990s concluded that in order to achieve better performance in the construction industry the priorities should be as follows:

1. Construction should meet the user requirements and be fit for the specific purpose for which it was commissioned

2. Consideration should be given to lower whole life costs through techniques such as value management
3. Clients should enjoy greater certainty over project time and costs
4. The industry should aim for greater efficiency through, for example, the elimination of waste in labour and materials.

The report identified that this could be achieved by:

*   Better integration of all stages in the construction process – design, planning and construction
*   Better management of construction supply chains – for example designers, subcontractors, specialist consultants, material suppliers
*   Much more consideration of end user requirements
*   Development of a learning culture on projects and within organisations
*   Greater consideration of the relationship between cost and value, whole life costs and the quality of contractors, and less reliance on lowest price
*   Longer-term relationships with clients and contractors to improve time, cost and quality
*   A move away from adversarial approaches
*   A better health and safety record.

Conclusion

The construction industry still persists in the practice of rewarding bad behaviour. If a contract is delayed, all participants get their money apart from the client, who has to pay! There can be no doubt that the pressure for change within the UK construction industry and its professions, including quantity surveying, is unstoppable, and that the volume of initiatives in both the public and private sectors to try to engineer change grows daily. The last decade of the twentieth century saw a realignment of the UK's economic base. Traditional manufacturing industries declined while services industries prospered, but throughout this period the construction industry has remained relatively static, with a turnover compared to GDP of 8%. The construction industry is still therefore a substantial and influential sector and a major force in the UK economy. Perhaps more than any other construction profession, quantity surveying has repeatedly demonstrated the ability to reinvent itself and adapt to change.

The remainder of this book will attempt to review the new opportunities that are presenting themselves to the quantity surveyor in a swiftly changing global construction market. It is not the object of this book to proclaim the demise of the traditional quantity surveyor practice offering traditional quantity services – these will continue to be in demand – but rather to outline the opportunities that are now available for quantity surveyors to move into a new era offering a range of services and developing new expertise that will truly make them 'The Masters of the Process'.

References

Auditor General (2000). *Modernising Construction.* HMSO.

Banwell, Sir H. (1964). *Report of the Committee on the Placing and Management of Contracts for Building and Civil Engineering Work.* HMSO.

Construction Industry Computing Association (2000) *Architectural IT Usage and Training Requirements.* http://www.cica.org.uk

CRINE (1994). *Cost Reduction Initiative for a New Era.* United Kingdom Offshore Operators Association.

Department of the Environment, Transport and the Regions (1998). *Rethinking Construction.* HMSO.

Edkins, A.J. and Winch, G.M. (1999) *Project Performance in Britain and France: the Case of Euroscan,* Barlett Research Paper 7.

Emmerson, Sir H. (1962). *Survey of Problems before the Construction Industries.* HMSO.

Goodall, J. (2000). *Is the British Construction Industry still suffering from the Napoleonic Wars?* Address to National Construction Creativity Club, London, 7 July 2000.

Latham, Sir M. (1994). *Constructing the Team.* HMSO.

Levene, Sir P. (1995) *Construction Procurement by Government.* An Efficiency Scrutiny, HMSO.

National Audit Office (2001) Modernising Construction, HMSO.

Property Eye Robinson *et al.* (Special Edition 2001). Project defects insurance. *Property Eye,* Sep.

Royal Institution of Chartered Surveyors (1971). *The Future Role of the Quantity Surveyor.* RICS.

Simon, Sir E. (1944). *The Placing and Management of Building Contracts.* HMSO.

Further reading

Agile Construction Initiative (1999). *Benchmarking the Government Client. Stage Two Study*. HMSO.

Building/MTI (1999). *QS Strategies 1999, Volumes 1 & 2*. Building/Market Tracking International.

Burnside, K. and Westcott. A. (1999). *Market Trends and Developments in QS Services*. RICS Research Foundation.

Cook, C. (1999). QS's in revolt. *Building*, 29 Oct, p. 24.

Department of the Environment, Transport and the Regions (2000). *KPI Report For The Minister for Construction*. HMSO.

Financial News (2000). Report finds majors shunning traditional work. *Building*, 24 Nov, p. 21.

Hoxley, M. (1998). *Value for Money? The Impact of Competitive Fee Tendering on the Construction Professional Service Quality*. RICS Research.

Pullen, L. (2001). What is best value in construction procurement? *Chartered Surveyor Monthly*, Feb.

Royal Institution of Chartered Surveyors (1993). *The Business of Building in France*. RICS.

Thompson, M. L. (1968). *Chartered Surveyors: The Growth of a Profession*. Routledge & Kegan Paul.

Watson, K. (2001). Building on shaky foundations. *Supply Management*, 23 Aug, pp. 23–26.

Web sites

Department of Environment Transport and the Regions – www.construction.detr.gov.uk

National Audit Office – http://www.nao.gov.uk

SIMAP – http://simap.eu.int

Treasury – http://www.hm-treasury.gov.uk

Tenders Electronic Daily (TED) – http://ted.eur-op.eu.int

Treasury Task Force – http://www.treasury-projects-taskforce.gov.uk/

2

Managing value.
Part 1: The supply chain

Introduction

This chapter examines the paths by which quantity surveyors can deliver a new range of added value services to clients, based upon increased client focus and a greater understanding of the function of built assets, including why new buildings are commissioned. Many clients who operate in highly competitive global markets base their procurement strategies on the amount of added value that can be demonstrated by a particular strategy. In order to meet these criteria quantity surveyors must: 'get inside the head' of their clients, fully appreciate their business objectives, and find new ways in which to deliver value and, conversely, remove waste from the procurement and construction process. The chapter will examine the application of manufacturing philosophies to the construction industry, and the role of the quantity surveyor in the implementation of these approaches.

Does your client feel good?

The dictionary definition of procurement is 'the management of obtaining goods and services'. For many years quantity surveyors took this to mean appointing a contractor who submitted the lowest (i.e. cheapest) tender price, based upon a bill of quantities and drawings, in competition with several other contractors. For public sector works a low priced tender was almost guaranteed to win the contract, as public entities had to be seen to be spending public money prudently and would have needed a very strong case to award the contract on any other basis. It is now clear that assembling an *ad hoc* list of six or so contractors, selected primarily on their availability, to tender for building work for which they have very little detailed information is not the best way to obtain value for money. In fact it does little more than reinforce the system in which contractors submit low initial prices secure in the knowledge that the contract will bring many opportunities to increase profit margins in the form of variation orders, claims for extensions of time/loss, and expense or mismanagement by the design team. Pre-Latham, the overriding ethos for procuring building works was to treat the supply chain with great suspicion; it was almost as if a Cold War existed between client, design team and contractor/subcontractor. The emphasis in construction procurement has now swung away from the system described above towards one that encourages partnerships and inclusion of the supply chain at an early stage – in fact, to a point where the definition of procurement could be restated as 'obtaining value for money deals.' As discussed in Chapter 1, there can now be few individuals involved within the construction process that do not believe that the design, procurement and construction of new built assets must become more efficient and client orientated. The evidence of wastage in terms of materials, time and money, not only in the short term but also throughout the lifecycle of a building, leaves the UK construction industry and its associated professions in an embarrassing position and open to criticism from all sides for participating in the production of such a low value product. Bernard Williams' study of *Building and Development Economics in the EC*, published by the *Financial Times* in 1993, demonstrated that the British construction industry was the least productive in Europe; as illustrated by the Euroscan project discussed in Chapter 1, this state of affairs appears still to exist. For whatever reason, the quantity surveyor – and the traditional brick-counting image enthusiastically fostered by so many within the industry, including the trade press – has also been the focus of this 'out of touch' image. For many years quantity surveyors have been seen as accountants to the construction industry, knights in shining armour, safeguarding clients to ensure that they receive a building as close as possible to the initial agreed target price, although in practice this has seldom been achieved. Traditionally a target cost has been set by the quantity surveyor in discussion with the client at the outset and then the process has been worked backwards, squeezing in turn the contractors, subcontractors and

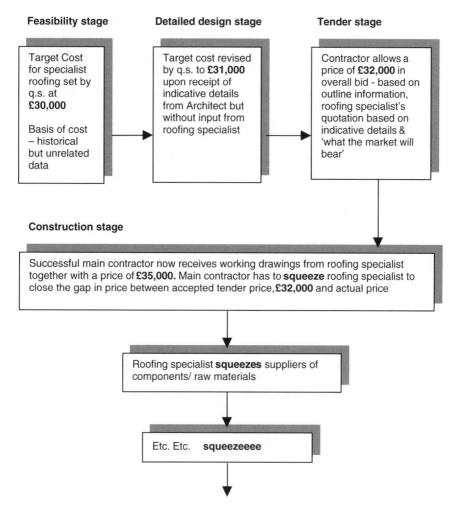

Figure 2.1 Supply chain squeeze.

suppliers in order to keep within this target cost. The squeezing increases in direct proportion to how far down the supply chain the organisation comes (see Figure 2.1).

The consequences of the supply chain squeeze illustrated in Figure 2.1 are low profit margins, lack of certainty and continuity for suppliers, delays in production, lack of consideration of whole life costs, suboptimal functionality, and rampant waste. In turn, low profits ensure that few (if any) resources can be channelled into research, technological improvement or quality assurance procedures. Construction productivity lags behind that of manufacturing, and yet manufacturing has been a reference point and a source of innovation in construction for many decades – for example, industrialised building, currently undergoing a resurgence in interest in the UK, and the use of computer-aided design come directly from the manufacturing sector. However, while some innovations have crossed the divide from manufacturing to construction, there has been little enthusiasm for other production philosophies. That said, new manufacturing industry-based approaches to supply chain management now being introduced into the construction industry enable the early involvement of suppliers and subcontractors in a project with devolved responsibility for design and production of a specific section of a building, with predicted and guaranteed whole life costs, for periods of up to 35 years. For clients, the benefits of this new approach include the delivery of increased functionality at reduced cost; for the supply chain members, benefits include certainty, less waste and increased profits.

Supply chain relationships and management

A construction project organisation is usually a temporary organisation designed and assembled for the purpose of a particular project. It is made up of different companies and practices that have not necessarily worked together before and are tied to the project by means of varying contractual arrangements. This is what has been termed a 'temporary multi-organisation'; its temporary nature extends to the workforce, which may be employed for a particular project, rather than permanently. These traditional design team/supply chain models are the result of managerial policy aimed at sequential execution and letting out the various parts of the work at apparently lowest costs. The problems for process control and improvement that this temporary multi-organisation approach produces are related to:

- Communicating data, knowledge and design solutions across the organisation
- Stimulating and accumulating improvement in processes that cross the organisational borders
- Achieving goal congruity across the project organisation

- Stimulating and accumulating improvement inside an organisation with a transient workforce.

The following quotation, attributed to Sir Denys Hinton when
he spoke at the RIBA in 1976, seems to sum up the traditional
attitude of building design teams:

> ... the so-called building team. As teams go, it really is rather
> peculiar, not at all like a cricket eleven, more a scratch
> bunch consisting of one batsman, one goalkeeper, a pole
> vaulter and a polo player. Normally brought together for a
> single enterprise, each member has different objectives,
> training and techniques and different rules. The relationship
> is unstable, and with very little functional cohesion and no
> loyalty to a common end beyond that of coming through
> unscathed.

Most of what is encompassed by the term 'supply chain management' was formerly referred to by other terms such as 'operations management', but the coining of a new term is more than
just new management speak; it reflects the significant changes
that have taken place across this sphere of activity. These
changes result from changes in the business environment. Most
manufacturing companies are only too aware of such changes –
increasing globalisation, savage price competition, increased
customer demand for enhanced quality and reliability etc.
Supply chain management was introduced in order that

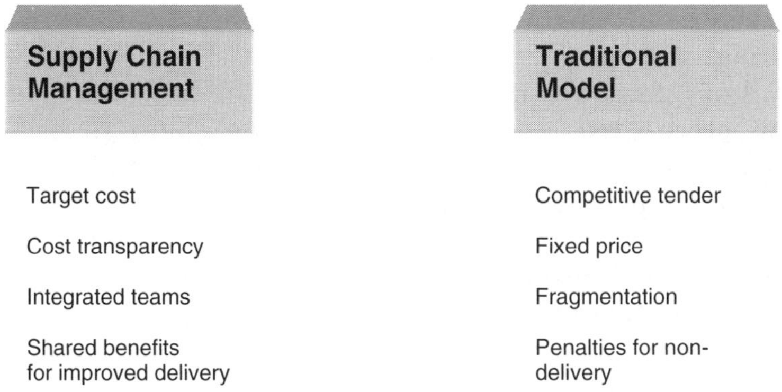

Supply Chain Management	Traditional Model
Target cost	Competitive tender
Cost transparency	Fixed price
Integrated teams	Fragmentation
Shared benefits for improved delivery	Penalties for non-delivery

Figure 2.2 Supply chain and traditional management approaches
compared.

manufacturing companies could increase their competitiveness in an increasingly global environment, and their market share and profits, by:

- Minimising the costs of production on a continuing basis
- Introducing new technologies
- Improving quality
- Concentrating on what they do best.

The contrast between traditional approaches and supply chain management is summarised in Figure 2.2.

The quantity surveyor as supply chain manager

What is the driving force for the introduction of supply chain management into the UK construction industry? The CRINE initiative in the oil industry (see Chapter 1) was the result of the collapse of world oil prices to $13 a barrel in 1992; however, in construction very little impetus has come from the industry and it is clients that are the driving force. Unlike other market sectors, because the majority of organisations working in construction are small, the industry has no single organisation to champion change (see Chapter 6). When Latham called for a 30% reduction in costs, the knee-jerk response from some quarters of the profession and industry was that cost = prices and therefore it was impossible to reduce the prices entered in the bill of quantities by this amount, and hence the target was unrealistic and unachievable. However, this was not what Latham was calling for. Reducing costs goes far beyond cutting the prices entered in the bill of quantities if it ever did; it extends to the reorganisation of the whole construction supply chain in order to eliminate waste and add value. The immediate implications of supply chain management are:

- Key suppliers are chosen on criteria, rather than job by job on competitive bids
- Key suppliers are appointed on a long-term basis and proactively managed
- All suppliers are expected to make sufficient profits to reinvest.

How many quantity surveyors have asked themselves, at the outset of a new project:

What does value mean for my client?

In other words, in the case of a new factory to manufacture, say, pharmaceutical products, what is the form of the built asset that will deliver value for money over the life cycle of the building for that particular client? For many years whenever clients have voiced their concerns about the deficiencies in the finished product, all too often the patronising response from the professions has been to accuse the complainants of a lack of understanding of design or the construction process, or both. The answer to the value question posed above will, of course, vary between clients; a large multinational manufacturing organisation will have a different view of value to that of a wealthy individual commissioning a new house, but it helps to illustrate the revolution in thinking and attitudes that must take place. In general, the definition of value for a client is 'a design to meet a functional requirement for a whole life cost'. Quantity surveyors are increasingly developing better client focus because only by knowing the ways in which a particular client perceives or even measures value (whether in a new factory or a new house) can the construction process ever hope to provide a product or service that matches these perceptions. Once these value criteria are acknowledged and understood, quantity surveyors have a number of techniques at their disposal in order to deliver a high degree of the feel good factor to their clients

Not all of the techniques are new. Many practising quantity surveyors would agree that the strength of the profession is expertise in measurement, and in supply chain management there is a lot to measure – for example:

- Productivity, for benchmarking purposes
- Value, to demonstrate added value
- Out-turn performance (not the starting point)
- Supply chain development (are the suppliers improving as expected?)
- Ultimate customer satisfaction (customers at a supermarket, passengers at an airport terminal, etc).

Of course, measuring value is extremely difficult to do. The Construction Industry Research and Information Association (CIRIA) and Loughborough University have been working with a clients' group, the Construction Round Table, to develop a client's guide for standardisation and pre-assembly. One of the tenets of this work is the need to measure success, and in particular to move this measurement beyond the obvious cash-related benefits. The team has developed specific 'measures' for both standardisation and pre-assembly, which are similar to the key performance indicators (KPIs) that are in use for other aspects in construction. Such generic measures are useful for industry-wide comparison, but have limited use in measuring project-specific issues. Therefore, in addition to these the team has produced methods of measuring benefits related to the drivers and constraints of the particular project. A related project called IMMPREST (Interactive Model for Measuring the benefits from Pre-assembly and Standardisation) is developing these tools further by comparing different methods of producing buildings or building elements against the following criteria:

- Monetary benefits
- Programme benefits
- Quality benefits
- Logistical and operational benefits
- Health and safety benefits
- Environmental and sustainability benefits
- Organisational benefits
- Other benefits.

What is a supply chain?

Before establishing a supply chain or supply chain network it is crucial to understand fully the concepts behind, and the possible components of, a complete and integrated supply chain. The term 'supply chain' is used to describe the sequence of processes and activities involved in the complete manufacturing and distribution cycle, including everything from product design through materials, component ordering, manufacturing and assembly to when the finished product is the hands of the final owner. Of course, the nature of the supply chain varies from industry to industry. Members of the supply chain can be

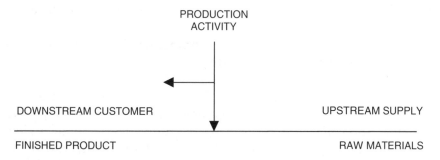

Figure 2.3 Supply chain.

referred to as upstream and downstream supply chain members (Figure 2.3). Supply chain management, which has been practised widely for many years in the manufacturing sector, therefore refers to how any particular manufacturer involved in a supply chain manages its relationship both up- and down-stream with suppliers to deliver cheaper, faster and better. In addition, good management means creating a safe commercial environment so that suppliers can share pricing and cost data with other supply team members.

The more efficient or lean the supply chain, the more value is added to the finished product. As if to emphasise the value point, some managers substitute the word 'value' for 'supply' to create the value chain. In a construction context, supply chain management involves looking beyond the building itself and into the process, components and materials that make up the building. Supply chain management can bring benefits to all involved when applied to the total process, which starts with a detailed definition of the client's business needs that can be provided through the use of value management and ends with the delivery of a building that provides the environment in which those business needs can be carried out with maximum efficiency and minimum maintenance and operating costs. In the traditional method of procurement, where the supply chain does not understand the underlying costs, suppliers are selected by cost and then squeezed to reduce the price and whittle away profit margins:

- Bids are based on designs to which suppliers have no input
- Low bids always win, but are unsustainable so costs are recovered by other means

- Margins are low, so there is no money to invest in development
- Suppliers are distant from the final customer, and therefore take limited interest in quality.

The traditional construction project supply chain can be described as a series of sequential operations by groups of people who have no concern about the other groups or the client and/or end user. In Figure 2.4 the end user is deliberately shown as being detached from the supply chain activity because typically no one thinks to consult this group, which may for example consist of patients in a health centre or passengers at an airport terminal, as to what they perceive a functionally efficient building to be. In rather more detail, the supply chain for, say, fenestration could be organised as shown in Figure 2.5.

The traditional supply chain arrangement (Figure 2.4) is characterised by lack of management, little understanding by tiers of other tiers' functions or processes, lack of communication, and a series of sequential operations by groups of people who have no concern about the other groups or client.

In supply chain management, however, prices are developed and agreed, subject to an maximum price with overheads, and

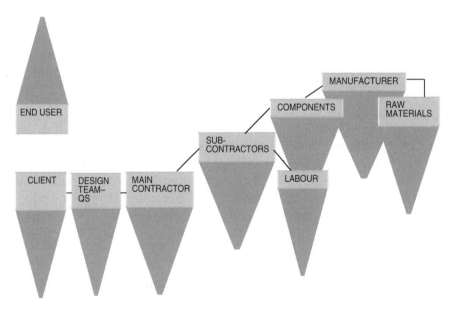

Figure 2.4 Traditional construction supply chain.

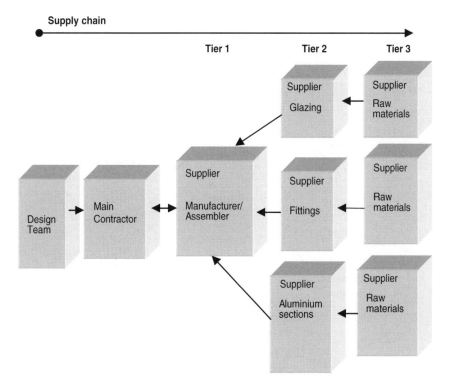

Figure 2.5 Fenestration supply chain.

profit is ring-fenced; all parties therefore collaborate to drive
down cost and enhance value with, for example, the use of an
incentive scheme. With the cost determined and profit ring-
fenced, waste can be attacked to bring down price and add value,
and there is an emphasis on continued improvement. Therefore,
in supply management:

* As suppliers account for 70–80% of building costs, they
  should be selected on their capability to deliver excellent
  work at competitive cost
* Suppliers should be able to contribute new ideas, products
  and processes
* Alliances are built outside the project
* Waste and inefficiency can be continuously identified and
  driven out.

Figure 2.6 shows the seven principles of supply chain management as suggested by the *Building Down Barriers Handbook*.

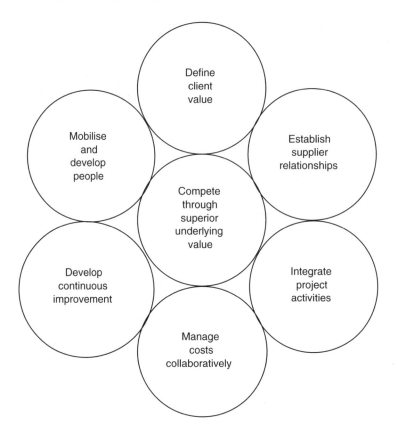

Figure 2.6 Principles of supply chain management (source: *The Building Down Barriers Handbook*, Defence Estates/BRE).

The philosophy of integrated supply chain management is based upon defining and delivering client value through established supplier links that are constantly reviewing their operation in order to improve efficiency. There are now growing pressures to introduce these production philosophies into construction, and it is quantity surveyors, with their traditional skills of cost advice and project management, who can be at the forefront of this new approach. For example, the philosophy of lean thinking, which is based on the concept of the elimination of waste from the production cycle, is of particular interest in the drive to deliver better value. In order to utilise lean thinking philosophy, the first hurdle that must be crossed is the idea that construction is a manufacturing industry that can only operate efficiently by means of a managed and integrated supply chain. At present the majority of clients are required to procure the design of a new building separately from the construction; however, as the subsequent delivery often involves a process where sometimes as much as 90% of the total cost of the completed building is delivered by the supply chain members, there would appear to be close comparisons with, say, the production of a motor car or an airplane.

The basics of supply chain management can be described as:

1. Determining which are the strategic suppliers, and concentrating on these key players as the partners who will maximise added value
2. Working with these key players to improve their contribution to added value
3. Designating these key suppliers as the 'first tier' on the supply chain and delegating to them the responsibility for the management of their own suppliers, the 'second tier' and beyond.

To give this a construction context, the responsibility for the design and execution of, for example, mechanical installations could be given to a first-tier engineering specialist. This specialist would in turn work with its second-tier suppliers, as well as with the design team, to produce the finished installation. Timing is crucial, as first-tier partners must be able to proceed in confidence that all other matters regarding the interface of the mechanical and engineering installation with the rest of the project have been resolved and that this element can proceed independently. At least one food retail organisation using supply chain management for the construction of its stores still places the emphasis on the tier partners to keep itself up to date with progress on the other tiers, as any other approach would be incompatible with the rapid timescales that are demanded.

Despite the fact that, on the face of it, certain aspects of the construction process appear to be prime candidates for this approach, the biggest obstacles to be overcome by the construction industry in adopting manufacturing industry-style supply chain management are:

1. That, unlike in manufacturing, the planning, design and procurement of a building is at present separated from its construction or production

2. The insistence that, unlike an airplane or motorcar, every build-
    ing is bespoke, a prototype, and therefore is unsuited to this
    type of model or for that matter any other generic production
    sector management technique. This factor manifests itself by:
    *   Geographical separation of sites, which causes breaks in
        the flow of production
    *   Discontinuous demand
    *   Working in the open air, exposed to the elements (can
        there by any other manufacturing process, apart from
        shipbuilding that does this?)
3. Reluctance by the design team to accept early input from
    suppliers and subcontractors, and unease with the blurring
    of traditional roles and responsibilities.

There is little doubt that the first and third hurdles result from
the historical baggage outlined in Chapter 1 and that, given time,
they can be overcome, whereas the second hurdle does seem to
have some validity despite statements from the proponents of
production techniques that buildings are not unique and common-
ality even between apparently differing building types is as high
as 70% (MoD, *Building Down Barriers*). Interestingly, though, one
of the main elements of supply chain management, just-in-time
(JIT), is reported to have started in the Japanese shipbuilding
industry in the mid-1960s – the very industry that opponents of
JIT in construction quote as an example for which, like construction, supply chain management techniques are inappropriate.
Therefore, discussion of the suitability of the application of supply
chain management techniques to building has to start with the
acceptance that construction is a manufacturing process that can
only operate efficiently by means of a managed and integrated
supply chain. One fact is undeniable – at present the majority of
clients are required to procure the design of a new building
separately from the construction. Until comparatively recently
international competition, which in manufacturing is a major
influencing factor, was relatively sparse in domestic construction
of major industrialised countries.

Adding value and minimising waste

One of the best-researched industries is car manufacturing.
Lean car production is characterised as using less of everything

compared with mass production – half the human effort, half the manufacturing space, and half the engineering hours to develop a new product in half the time. The competitive benefits created by means of the new approach seem to be remarkably sustainable; however, with the exception of quality methodologies, this new philosophy is little known in construction. The ideas of the new production philosophy first originated in Japan in the 1950s. The most prominent application, as pioneered by Taiichi Ohno (1912–1990), was the Toyota production system, and central to the system was the single-minded determination to eliminate waste through, among other techniques, co-operation with suppliers. Ohno was credited with the identification of waste, defined as 'activities that create no value' – *defects* in products, *overproduction* of goods not needed, *inventories* of goods awaiting further processing or consumption, unnecessary *processing*, unnecessary *movement* of people, unnecessary *transport* of goods and *waiting* by downstream activities for processes to finish on an upstream activity. In 1996 Womack and Jones added an eighth; the design of goods and services *unsuitable* for users' needs. How many of the items highlighted above can be identified in a typical construction project – four? Five? More?

Simultaneously, under the guidance of American consultants, quality issues became another focus for Japanese industry. By the beginning of the 1990s the new production philosophy, known by several different names, including world class manufacturing, the new production system and lean production, was in widespread use. The term 'lean construction' was introduced by DETR (1998), and central to this new direction are the concepts of just-in-time (JIT) and total quality control (TQC). Many other related concepts, too numerous to describe here, have been added to the original lean production philosophy; however, three are included as particularly relevant for the quantity surveyor in the drive for added value in construction – value engineering/management, concurrent engineering, and continuous improvement. The core of the new production philosophy is in the observation that there are two kinds of phenomena in all new production systems:

1. Conversions
2. Flows.

Conversions versus flows

The conventional approach to a construction project based on conversions is illustrated in Figure 2.7.

The conversion approach to construction revolves around the principle that buildings are conceived as sets of operations controlled individually for the least cost. However straightforward the conversion appears, it masks (particularly in the construction process) inherent waste owing to:

• Rework due to design or construction errors
• Non-value-adding services in the material and work flows, such as waiting and handling and double handling
• Inspecting, duplicating activities and accidents.

The primary focus in the design of new built assets is therefore on minimising value loss, whereas in construction it is on minimising waste.

The following statistics are taken from various reports from CIRIA and Movement for Innovation:

• Every year in the UK approximately 13 million tonnes of construction materials are delivered to site and thrown away, unused

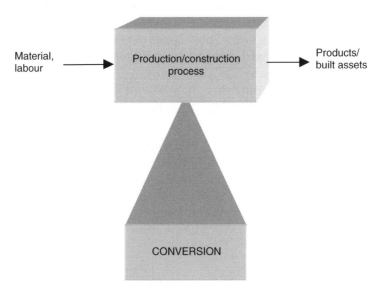

Figure 2.7 Conversion process.

- 10% of products are wasted through oversupply, costing £2.4 billion per annum
- Another £2.4 billion per annum is wasted in stockpiling materials
- £5 billion per annum is squandered by the misuse of materials.

The lean construction tool box

The following approaches are available to achieve lean construction:

1. Just-in-time production
2. Total quality control/continuous improvement
3. Value analysis, engineering and management
4. Concurrent engineering
5. Prime contracting.

Just-in-time production

Just-in-time (JIT) production was the starting point for Ohno and the Toyota revolution. This so-called pull-type production process is based on the principle that good communications with the suppliers ensure that production is initiated by actual demand rather than by plans based on forecasts. It contrasts with the push-type technology, where large volumes of materials and components are produced, transported and stored ahead of demand and requirements. The driving idea is the reduction or elimination of work in progress – large stocks of goods and materials that have been produced and are awaiting further processing and/or transportation before completion – thereby eliminating two or possibly three of Ohno's causes of waste.

Delays on site can often be the result of the unavailability of materials. For example, a quantity surveyor has prepared a schedule of the door and window requirements for a new hospital, which has been designed by a firm of consultant architects, but without reference to material and component availability. The schedule becomes part of the successful contractor's bid, and as work progresses the contractor submits an order for 1500 hardwood double-glazed window units, 2400 × 1200 mm. The

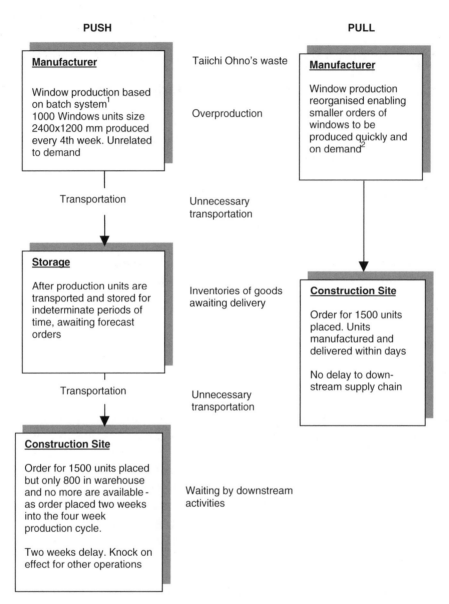

Figure 2.8 Lean production of building components.

different response paths to this order are illustrated in Figure 2.8. The first path reflects the traditional approach to production, the so-called push method, where the supplier is not able to fill the order because of lack of units and inflexible pre-set production schedules. The second path reflects the pull method, which is able to react more quickly to demand. In the centre of Figure 2.8 is a list of some of the elements of waste identified by Taiichi Ohno; each time waste is removed from the process, value is added! This example illustrates the potential problems with just one component of a building project and how, traditionally, many of the components commonly used in construction have large elements of waste built into their production. However, this sort of scenario can be encountered many times during the construction of a project.

Every time this waste is removed from the supply chain, value is added to the process, leading to lower costs, shorter construction periods and greater profits. Perhaps some of the statistics of time and material wastage referred to earlier could be improved with the wide-scale adoption of JIT techniques and the realisation by the design team that they have to design and manage a project accordingly.

Total quality control/continuous improvement

The starting point of the quality movement is the inspection of raw materials and products using statistical methods. Quality methodologies have developed in parallel with the evolution of the concept of quality. The focus has changed from inspection, through process control to continuous process improvement. In the traditional approach to quality, no special effort is made to eliminate defects, errors or omissions or to reduce their impact. In numerous studies from different countries, the cost of poor quality has been found to be between 10 and 20% of total project costs. Agreements for the procurement of construction services by means of a supply chain should include:

- The level of service required stated in performance or output terms
- The consequences of failure
- The means by which the client is able to measure the supply chain's performance.

Under the contract, there must be a mechanism that enables monitoring of the performance of the supply chain. This could be in the form of self-monitoring, subject to periodic checks by the client; open book accounting is normally required for this to work effectively. In Chapter 1 the process of benchmarking was described, but what should be benchmarked and how? One approach is to introduce risk–reward schemes that enable the supply chain to share in the benefit if the works are completed under the target cost; conversely, the supply chain will share the client's downside if the final cost exceeds the target. This approach differs from the establishment of a guaranteed maximum price, and this is thought by some practitioners to be at odds with the collaborative team approach.

Figure 2.9 illustrates such a scheme based upon the target cost, in which the dead band has been agreed beforehand at £1 million. This band is not fixed, and will reflect the confidence of the parties in the target value of a particular project. Similar arrangements can be made for completion targets, as illustrated in Figure 2.10.

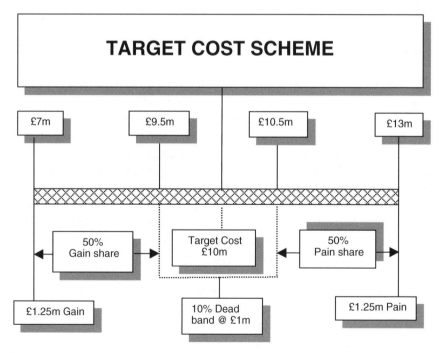

Figure 2.9 Target cost scheme.

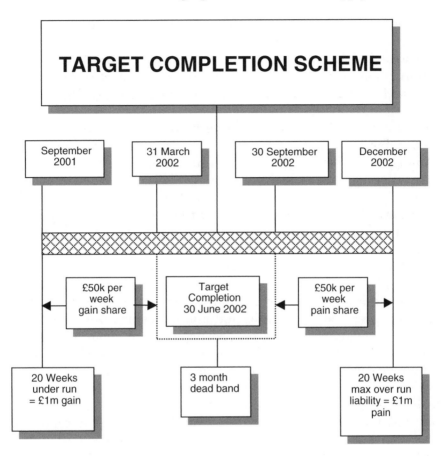

Figure 2.10 Target completion scheme.

Value analysis, engineering and management

The techniques of value engineering and management, originally known as value analysis, are central to the goal of delivering built assets that meet the functional and operational needs of a client. Developed in the USA by Lawrence D. Miles in the immediate post-Second World War era for the manufacturing and production sectors, value engineering and management are now widely practised by UK quantity surveyors in both public and private sectors. To quote Robert N. Harvey, one-time manager of capital programmes and value management for the Port Authority of New York and New Jersey: 'Value engineering is like love – until you've experienced it you just can't begin to understand it'. In the early 1990s the Port Authority conducted value engineering workshops on nearly $1 billion worth of construction projects. The total cost of the workshops was approximately $1 million, a massive statement of confidence in the technique, which paid off by delivering nearly $55 million in potential savings.

For a somewhat more objective view of the process, perhaps the reference point should be the International Society of American Value Engineers' definition of value engineering as:

> A powerful problem solving tool that can reduce costs while maintaining or improving performance and quality. It is a function-oriented, systematic team approach to providing value in a product or service.

The philosophy of value engineering and management is a step change from the traditional quantity surveying belief that delivering value is based on the principle of slashing costs to keep within the original budget – what was and still is euphemistically referred to as 'cost reconciliation'. The basis of value management is to analyse, at the outset, the function of a building (or even part of a building) as defined by the client or end user, and then, by the adoption of a structured and systematic approach, to seek alternatives and remove or substitute items that do not contribute to the efficient delivery of this function – thereby adding value. The golden rule of value engineering/management is that, as a result of the value process the function(s) of the object of the study should be maintained and if possible enhanced, but never diminished or compromised. Therefore, once again, the focus for the production of the built asset is a client's perception of value. Perhaps before continuing much further the terms associated with various value methodologies should be explained. The terms in common usage are value analysis, value engineering and value management.

Value analysis

Value analysis was the term adopted by Lawrence D. Miles for his early studies, and is defined as 'an organised approach to the identification and elimination of unnecessary cost'.

Value engineering

Value engineering was the term adopted in 1959 by the International Society of American Value Engineers (SAVE) when it was established to formalise the Miles approach. The term is widely used in North America, and its essential philosophy is 'a disciplined procedure directed towards the achievement of necessary function for minimum cost without detriment to quality, reliability, performance or delivery'. As if to emphasise the importance placed on value engineering in 2000, Property Advisors to the Civil Estate (PACE) introduced an amendment to GC/Works/1 – Value Engineering Clause 40(4). The amendment states:

> The Contractor shall carry out value engineering appraisals throughout the design and the construction of the Works to identify the function of the relevant building components and to provide the necessary function reliability at the lowest possible costs. If the Contractor considers that a change in the Employer's Requirements could effect savings, the Contractor shall produce a value engineering report.

Value management

Value management involves considerably more emphasis on problem solving, as well exploring in-depth functional analysis and the relationship between function and cost. It also incorporates a broader appreciation of the connection between a client's corporate strategy and the strategic management of the project. In essence, value management is concerned with the 'what' rather than the 'how', and would seem to represent the more holistic approach now being demanded by some UK construction industry clients – i.e. to manage value. The function of value management is to reduce total whole life costs, comprising initial construction costs, annual operating, maintenance and energy costs, and periodic replacement costs, without adversely affecting and while indeed improving performance and reliability and other required design parameters. It is a function oriented study and is accomplished by evaluating functions of the project and its subsystems and components to determine alternative means of accomplishing these functions at lower cost. Using value management, improved value may be delivered in three predominant manners:

1. By providing for all required functions, but at a lower cost
2. By providing enhanced functions at the same cost
3. By providing improved function at a lower cost – the Holy Grail.

Among other techniques, value management uses a value engineering study or workshop that brings together a multidisciplinary team of people, independent of the design team, who own the problem under scrutiny and have the expertise to identify and solve it. A value engineering study team works under the direction of a facilitator, who follows an established set of procedures (for example the SAVE value methodology standard, as in Figure 2.11) to review the project, making sure the team understands the client's requirements, and develops a cost-effective solution. Perhaps the key player in a value engineering study is the facilitator or value management practitioner, who must within a comparatively short time ensure that a group of people work effectively together. An example is Alphonse Dell'Isola, the Washington DC-based practitioner, who has risen to be an icon in value management circles and has helped SAVE to prove their claim that value management produces savings of 30% of the estimated cost for constructing a project and that for every pound invested in a value engineering study, including participants' time and implementation costs, £10 is saved. Certainly, organisations that have introduced value engineering into their existing procurement process (e.g. previously publicly-owned water companies, London Underground, etc.) all report initial savings of around 10–20%. In some respects, value management is no more than the application of the standard problem solving approach to building design. If there is one characteristic that makes value engineering/value management distinctive, it is the emphasis given to functional analysis.

The techniques that can be used to define and analyse function are:

• Value trees
• Decision analysis matrices
• Functional analysis system technique (FAST) diagrams
• Criteria scoring.

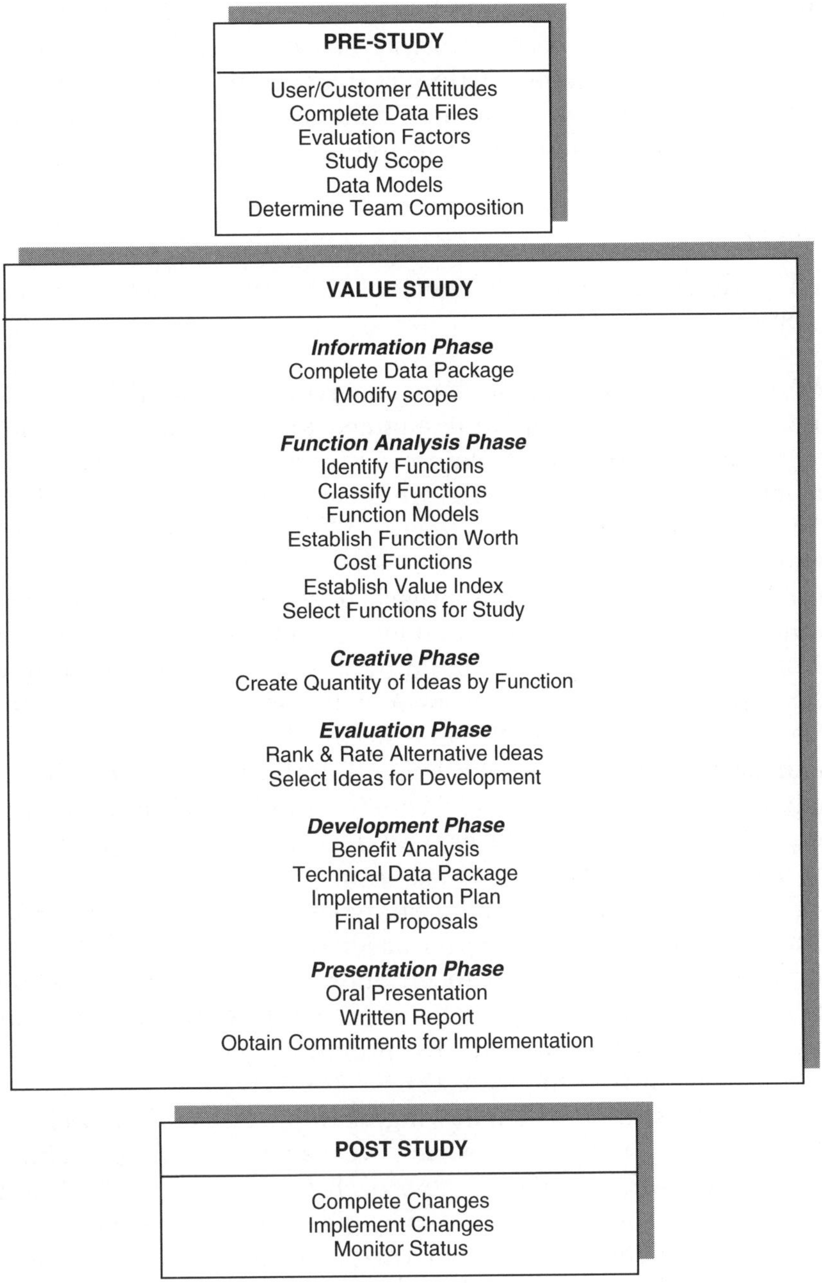

PRE-STUDY

User/Customer Attitudes
Complete Data Files
Evaluation Factors
Study Scope
Data Models
Determine Team Composition

VALUE STUDY

Information Phase
Complete Data Package
Modify scope

Function Analysis Phase
Identify Functions
Classify Functions
Function Models
Establish Function Worth
Cost Functions
Establish Value Index
Select Functions for Study

Creative Phase
Create Quantity of Ideas by Function

Evaluation Phase
Rank & Rate Alternative Ideas
Select Ideas for Development

Development Phase
Benefit Analysis
Technical Data Package
Implementation Plan
Final Proposals

Presentation Phase
Oral Presentation
Written Report
Obtain Commitments for Implementation

POST STUDY

Complete Changes
Implement Changes
Monitor Status

Figure 2.11 Value management methodology (source: SAVE International Society of American Value Engineers).

Once the function of an item has been defined, the cost or worth can be calculated and the worth/cost ratio scrutinised to determine value for money. Value management therefore can be described as a holistic approach to managing value that includes the use of value engineering techniques.

The process

The theory of value management is – buy function, don't buy product.

While it is not the purpose of this book to give a detailed description of every stage of a value engineering workshop, it is worth spending some time to explain the process as well as examining in more detail the functional analysis phase (see Figure 2.11). Value management has its roots in the manufacturing sector, where it has been around for many years, and there can be problems in applying the approach to the construction of a new building. Nevertheless, valuable insights into the functions behind the need for a new building and what is needed to fulfil these functions can flow from a value engineering workshop, even if it is not as lengthy and detailed as the traditional 40-hour programme used in the USA. If nothing else, it may be the only time in the planning and construction of a project when all the parties – client, end user, architect and quantity surveyor – sit down together to discuss the project in detail.

For example, let's consider a hypothetical project where a value engineering workshop is used: the construction of a new clinic for the treatment of cancer. There are numerous variations on and adaptations of both the approach to conducting a value engineering workshop and the preparation of items like the FAST diagram (see Figure 2.12).

As an illustration, the following example is based on a classic 40-hour, 5-day value-engineering workshop, as this presents a more proven and pragmatic approach than some of the latter-day variations of value engineering. The workshop team is made up of six to eight experts from various design and construction disciplines who are not affiliated to the project, as it has been found that the process is not so vigorous if in-house personnel are used. In addition, an independent facilitator is recommended, again because he or she will be less liable to compromise on the delivery of any recommendations. The assembled

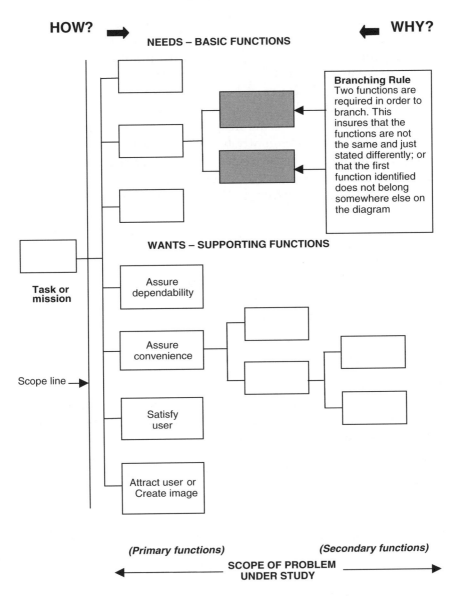

Figure 2.12 FAST methodology (source: SAVE International Society of American Value Engineers).

team then commences the workshop, following the steps of the SAVE methodology (Figure 2.11). At the start of the week the group is briefed on the project by the clinic personnel and members of the design and construction team, and the scope of the study is defined. Costs of the project are also carefully examined and analysed, using a variety of techniques, as well as being compared to other facilities with a similar function.

The first major task in the study is the functional analysis phase, during which the most beneficial areas for value improvement will be identified. Whilst unnecessary cost removal has been the traditional target for quantity surveyors, it is important to emphasise that more frequently today value studies are conducted to improve a building performance without increasing cost – or, to express it more simplistically, to maximise 'bang for your buck'.

Functional analysis using a FAST model follows the following steps:

* Defining function
* Classifying function
* Developing functional relationships
* Assigning cost to function
* Establishing function worth.

Defining function

Defining function can be problematic; experience has shown that the search for a definition can result in lengthy descriptions that do not lend themselves to analysis. In addition, the definition of function has to be measurable. Therefore, a method has been devised to keep the expression of a function as simple as possible; it is a two word description made up from a verb and a

Table 2.1 Typical verbs and nouns used in functional analysis

Verbs		Nouns	
Amplify	Limit	Area	Power
Attract	Locate	Corrosion	Protection
Change	Modulate	Current	Radiation
Collect	Move	Damage	Repair
Conduct	Protect	Density	Stability
Contain	Remove	Energy	Surface
Control	Rotate	Flow	Vibration
Enclose	Secure	Fluid	Voltage
Filter	Shield	Heat	Volume
Hold	Support	Insulation	Weight

noun. Table 2.1 shows two lists of typical verbs and nouns; more comprehensive lists are readily available.

On first sight this approach may appear to be contrived, but it has proved effective in pinpointing functions. It is not cluttered with superfluous information and promotes full understanding by all members of the team, regardless of their knowledge or technical backgrounds. For example: identify treatment, assess condition, diagnose illness (Figure 2.13).

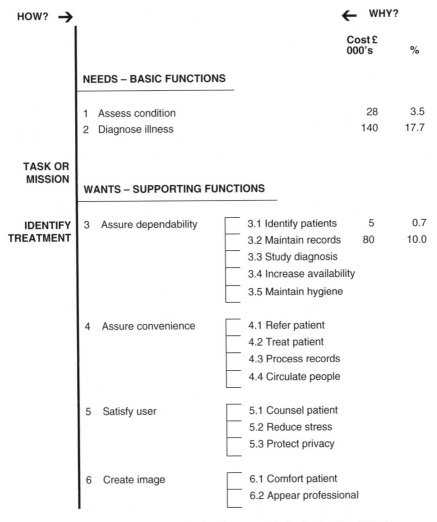

Figure 2.13 Functional Analysis System Technique (FAST) diagram

Classifying function

In order to establish some sort of hierarchy, functions are classified into primary or basic functions, and supporting functions. Basic functions or needs are functions that make the project or service work; if they are omitted this will have an impact on the effectiveness of the completed project. Out of the list of basic functions emerges the highest order function, which can be defined as the overall reason for the project, and meets the overall needs of the client. This function is placed to the left of the scope line on the FAST diagram (see Figures 2.12 and 2.13). Supporting functions, in the majority of cases, contribute nothing to the value of a building, and generally are classified as:

* Assure dependability
* Assure convenience
* Satisfy the user
* Create an image.

At first glance these categories may seem to have little relevance to construction related activities, until it is understood that, for example, the 'create an image' heading includes items such as aesthetic aspects, overall appearance, decoration and implied performance (reliability, safety, etc.). Items that in themselves are not vital for the integrity of the project may nevertheless be high on the client's list of priorities.

Developing functional relationships

Functional analysis system technique (FAST) models provide a method of depicting functional relationships (see Figures 2.12 and 2.13). The model works both vertically and horizontally by first determining the highest order function, called the task or mission, which is positioned to the left of the vertical scope line. By using the verb/noun combination and working from the left asking the question 'HOW?' and from the right asking the question 'WHY?', the functions and their inter-relationships can be mapped and their value allocated at a later phase.

Assigning cost to function

Conventionally project costs are given in a detailed cost plan, where the actual costs of labour materials and plant are calculated and shown against an element, as in the Building

Table 2.2 Elemental costs (Building Cost Information Service)

Element	Total cost of element (£)	Cost per m² of gross floor area (£)	Element unit quantity	Element unit rate (£)
2G Internal walls and partitions	430 283	45.00	8025 m²	53.62

Cost Information Service's standard list (for example, see Table 2.2).

Value engineering is based on the concept that clients buy functions (not materials or building components) as defined and expressed by their user requirements. Therefore, splitting costs among the identified functions shows how resources are spent in order to fulfil these functions. Costs can then be viewed from the perspective of how efficiently they deliver the function. Obviously the cost of each element can cover several functions – for example, the element BCIS Ref 2G Internal Walls and Partitions may contribute to the delivery of several functions of the project. It is therefore necessary at the outset to study the cost plan and to allocate the costs to the appropriate function (see Table 2.3).

Table 2.3 Cost plan allocation to functions

Element: 2G Internal Walls and Partitions		Elemental Cost: £430 283
	Function	Cost (£)
3.1	Identify patients	12 000
3.2	Maintain records	30 000
3.3	Study diagnosis	25 900
3.4	Increase availability	6 900
3.5	Maintain hygiene	7 000
4.1	Refer patient	45 000
4.2	Treat patient	56 000
4.3	Process records	40 000
4.4	Circulate people	60 889
5.1	Counsel patient	26 605
5.2	Reduce stress	4 989
5.3	Protect privacy	38 000
6.1	Comfort patient	34 000
6.2	Appear professional	43 000
		£430 283

A similar exercise is carried out until all of the project costs are allocated to functions.

Establishing function worth

The next step is to identify which of the functions contains a value mismatch, or in other words seems to have a high contribution to the total project cost in relation to the function that it performs. The creative phase will then concentrate on these functions. Worth is defined as ' the lowest overall cost to perform a function without regard to criteria or codes.' Having established the worth and the cost, the value index can be calculated using the formula: value = worth/cost. The benchmark is to achieve a ratio of 1.

The FAST diagram illustrated in Figure 2.13 is characterised by the following:

- The vertical 'scope line', which separates and identifies the highest level function (the task or mission) from the basic and supporting functions. It is pivotal to the success of a functional analysis diagram that this definition accurately reflects the mission of the project.
- The division of functions into Needs or Basic functions, without which the project will not meet client requirements, and Wants or Supporting functions (usually divided into the four groups discussed above), without which the project could still meet the client's functional requirements.
- The use of verb/noun combinations to describe functions.
- Reading the diagram from the left and asking the question 'how is the function fulfilled?' provides the solution.
- Reading the diagram from the right and asking the question 'why?' identifies the need for a particular function.
- The right-hand side of the diagram allows the opportunity to allocate the cost of fulfilling the functions in terms of cost and percentage of total cost.

Therefore, the FAST diagram in Figure 2.13 clearly shows the required identified functions of the project, together with the cost of providing those functions. What follows is the meat of the workshop – a creative session that relies on good classic brainstorming of ideas, a process that has been compared by those who have experienced it to a group encounter session, and the aim of which is to seek alternatives. The discussion may be structured or unstructured – Larry Miles was quoted as saying that the best atmosphere in which to conduct a study was one laced with cigarette smoke and Bourbon, but in these more politically correct times these aids to creativity are seldom employed! The rules are simple. Nobody is allowed to say, 'that won't work'. Anybody can come up with a crazy idea. These sessions can generate hundreds of ideas, of which perhaps 50 will be studied further in the workshop's evaluation phase. Those ideas will be revisited, and some discussion will take place as to their practicality and value to the client. Every project will have a different agenda. The best of the recommendations are then fully developed by the team, typically on day four of the workshop, and studies are carried out into costs and whole life costs of a proposed change before presentation to the client on the final day. It is an unfortunate fact of life of the classic 5-day workshop that the team member tasked with costing the recommendations has to work into the night on the penultimate day. Ultimately a draft report is approved, and a final report is written by the team leader. In addition to the above procedures, risk assessment can (or, according to opinion, should) be introduced into the process. As the value analysts go through and develop value recommendations they can be asked to identify risks associated with those recommendations, which can either be quantitative or qualitative. And if brainstorming sounds just a little esoteric for the quantity surveying psyche, take heart; the results of a value engineering workshop usually produce tangible results that clearly set out the costs and recommendations in a very precise format (Figure 2.14).

The question is often asked, are there buildings that are beyond value management? The answer is most certainly – yes. There are many high-profile examples that flaunt the drive to lean construction, and these mainly fall into the category of projects that have the highest order function of making a statement – commercially, politically or otherwise. L'Arche Défense, constructed to begin the revitalisation of La Défense, the futuristic Parisian business quarter, and of course Pei's glass pyramid at the Louvre, are good examples of such projects.

PROJECT Cancer Treatment and Research Clinic	VALUE ENGINEERING PROPOSAL
PROPOSAL Eliminate return duct to ventilation system	DATE
	ITEM No H14

ORIGINAL PROPOSAL:

Each room has a return grille and ductwork connecting back to a return fan.

PROPOSED CHANGE:

Eliminate duct return system on individual floors and provide an above ceiling return plenum.

ADVANTAGES:

More available ceiling space
Balancing of return system is simplified

DISADVANTAGES:

Plenum rated cable, tubing and pipe required
May be acoustic transmission problems in walls

COST SUMMARY	INITIAL COST	OPERATION AND MAINTENANCE COST – 15 Years		TOTAL LIFE CYCLE COST
		PER ANNUM	LIFE CYCLE – PV @ 6%	
ORIGINAL PROPOSAL	£149,450	£4,000	£38,848	£188,298
VE PROPOSAL	£86,000	£2,000	£19,424	£105,424

Figure 2.14 Costs and recommendations of value engineering proposal.

Concurrent engineering

Concurrent or simultaneous engineering deals primarily with the detail design stage. The term refers to an improved design process characterised by rigorous upfront functional analysis, incorporating the constraints of subsequent phases into the conceptual phase, in contrast to the traditional sequential design process.

There can be little argument that the greatest improvements through the implementation of supply chain management techniques, including JIT, have been achieved in industries where high volume, repetitive manufacture is the norm. In fact, some commentators have suggested that truly successful implementation can only be attained in Japan because of the unique cultural factors and employee/employer relationships that exist there. Whether supply chain management techniques are suitable for every project is still a matter of opinion; the system is not a panacea for problems of every project, from small one-off projects to large multi-million pound infrastructure schemes. However, if only some of the processes described above are introduced, then at least this is a start to focusing on delivering client orientated building solutions that function efficiently. From other industries that have adopted value chain management comes the warning – be patient. Lean production is not a quick fix, and major changes in mindset and skills take time: at least 1–2 years for basic understanding, another 3–4 years for training, and 2–4 years to achieve sustaining skills and behaviours. The various commercial web-enabled management tools that permit real time communication between the various supply chain members are powerful implements in the successful operation of supply chain management (see Chapter 5). To some within the industry, a logical step from the wide scale use of supply chain management techniques is prime contracting – or, as it is sometimes known, the 'one stop shop' approach to procurement.

Prime contracting

Perhaps the biggest step change in the attitude of clients towards the commissioning of new buildings comes with the emergence of prime contracting, which is now in use with influential clients in both public (e.g. Ministry of Defence) and private (e.g. J. Sainsbury) sectors. Prime contracting relies heavily on the use of integrated supply chain management, and the philosophy of prime contracting can be described as the search for continuous improvements in time, cost, safety, and quality and whole life cost considerations by means of collaborative working. A prime contractor can be defined as one having the overall responsibility for the management and delivery of a project, including co-ordinating and integrating the activities of a number of subcontractors to meet the overall specification efficiently, economically and on time. For the client, the major attraction is that the traditional dysfunctional system is replaced by single-point responsibility. The prime contractor has the responsibility for:

- Total delivery of the project in line with whole life predictions, which can be up to 35 years from the time of delivery
- Subcontractor/supplier selection – note there are exceptions to this rule, for example where clients may, because of their influence or market position, be able to procure some items more cheaply than via the existing supply chain
- Procurement management
- Planning, programming and cost control
- Design co-ordination and the overall system engineering and testing.

Considering the major organizations currently using prime contracting, while the Ministry of Defence requires contractors to include maintenance over the lifecycle of the project/contract in the deal, J. Sainsbury restricts the inclusion of whole life costs to an occasional audit. This is because the ever-changing demands of J. Sainsbury's business requirements do not, at present, lend themselves to long-term (e.g. 25-year) life cycle cost considerations. The client generally has one point of contact, the prime contractor (Figure 2.15), and this arrangement replaces the traditional situation where a client is confronted with a series of relationships that can be categorised into contractual and procedural, often resulting in a complex mesh of relationships that seem designed to provide a smokescreen to hide inefficiency and accountability. This is the equivalent of a new car buyer having to procure the engine, body shell and gear box from independent sources.

Many clients, in both the public and private sectors, do not wish to contract with a whole range of different contractors, designers, suppliers and subcontractors. Clients talk about single-point supply delivering them peace of mind, with assurances as to time, cost and quality from a robust prime contractor

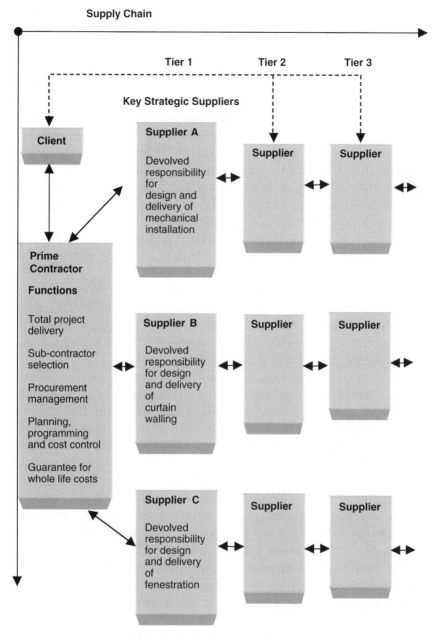

- - - - - Step in agreement for client in areas of commercial benefit; e.g. – a supermarket chain's influence in the purchase of refrigeration equipment and plant

Figure 2.15 Prime contracting using supply chain management.

able to manage and harness effectively the value stream outputs from the supply chain. But who can be the single-point supplier – does it have to be a large contractor with a correspondingly large balance sheet? The answer, in theory, is no; provided the particular requirements of the client match the outputs channelled through the single-point supplier, and assuming that such a supplier has the management skills, there is nothing to stop a firm of consultants acting as prime contractor, although to date only one such consultancy firm has risen to the challenge. Prime contractors take full responsibility for the performance of their subcontractors and consultants; however, it is also necessary for the prime contractor to demonstrate an ability to bring together all the parties of the supply chain. This could be achieved either by entering into partnering style alliances, or by a series of non-binding partnering protocols (see Chapter 3).

The Ministry of Defence uses an approach to prime contracting where it develops so-called 'clusters' or elements of the supply chain that constitute an integrated team, and these work together for a particular part of the installation – for example, groundworks, lift installation, roofing, etc.

These clusters, or tiers, are built outside of particular projects, and there could be two or three supply chains capable of delivering an outcome for each cluster. A typical cluster for, say, mechanical and electrical services could include the design engineers, the contractor and the principal component manufacturers. Crucially, for the success of this approach, clusters must have the confidence to proceed in the design and production of their element in the knowledge that clashes in design or product development with other clusters are being managed and avoided by the prime contractor. Without this assurance this approach offers little more than the traditional supply chain management techniques, where abortive and unco-ordinated work is unfortunately the norm. It should be noted that the legal structure of such clusters has yet to be formalised.

Supply chains are unique, but it is possible to classify them generally by their stability or uncertainty on both the supply side and the demand side (Figure 2.16).

On the supply side, low uncertainty refers to stable processes while high uncertainty refers to processes that are rapidly changing or highly volatile. On the demand side, low uncertainty

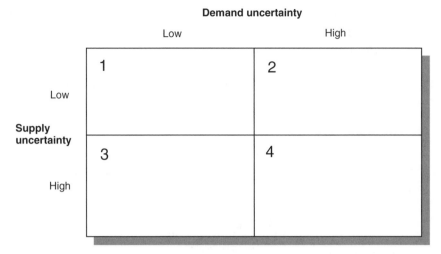

Figure 2.16 Supply and demand uncertainty.

relates to functional products in a mature phase of the production lifecycle while high uncertainty relates to innovative products. Once the chain has been catagorised, the most appropriate tools for improvement can be selected.

Another important step in creating the right environment for prime contracting is the approach towards liability for defects. In the traditional approach the various parties make their own arrangements for indemnity insurance, whereas in some prime contracting models blanket professional indemnity insurance cover is arranged for the whole supply chain, thereby eliminating the blame culture that pervades the industry and reinforcing the sense of well being for the client. This approach theoretically also reduces the overall costs of insurance.

To date, the role of the quantity surveyor in prime contracting has tended to be as works advisor to the client – a role that includes examination of the information produced by the prime contractor (and as such resented by prime contractors) and being cost consultant to a prime contractor, without the necessary in-house disciplines. However, there is nothing to prevent a quantity surveying consultancy from taking the prime contractor role, provided the particular requirements of the client match the outputs and that added value can be demonstrated.

Several models of prime contracting procurement are being developed. Figure 2.17 illustrates the path used by the Ministry of Defence. The approach is characterised by:

- The early involvement of the prime contractor in the process
- The selection of the prime contractor during the second stage of the process, along the lines of a public–private partnership deal
- The devolution of responsibility for the complete design, execution and delivery of the project with guaranteed whole life costs.

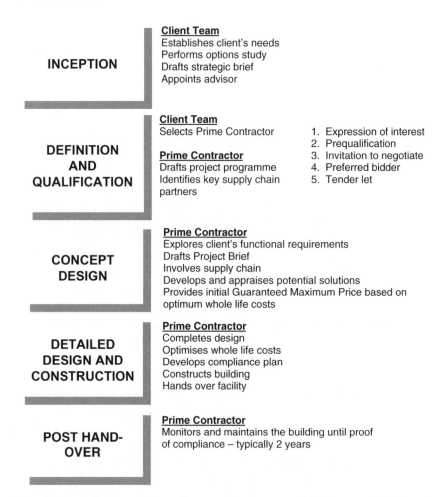

Figure 2.17 Prime contracting procurement path. Source: Holti, R. *et al.* Prime Contractor Handbook of Supply Chain Management. Defence Estates.

Conclusion

The government's contract guidance for public construction, announced in May 2000, requires all central government clients to limit their procurement strategies for the delivery of new works from 1 June 2000, and for refurbishment and maintenance contracts from 1 June 2002, to PPPs, design and construct, and prime contracting, unless it can be clearly shown that a traditional, non-integrated strategy offers best value for money. In announcing this policy, the Chief Secretary to the Treasury stated that:

> The Achieving Excellence initiative, launched last year to improve significant Government clients' performance, made it clear that we will focus on interacting with suppliers in the future through integrated supply chains working co-operatively.

In the private sector, too, the Movement for Innovation (M4i) and the Design Building Foundation (DBF) are continuing to drive forward the message about the integration of supply chain management.

References

DETR (1998). *Rethinking Construction*. Report of the Construction Task Force.

Further Reading

Bryant, J. W. (1998). *Monograph, Function: Definition and Analysis*. SAVE International.

Cnudde, M. (1991). Lack of quality in construction – economic losses. *Proceedings of the European Symposium on Management, Quality and Economics in Housing and Other Building Sectors, Lisbon, September 30–October 4*.

Dell'Isola, A. (1997). *Value Engineering; Practical Applications*. R. S. Means Co.

Fisher, M. (1997). What is the right supply chain for you? *Harvard Business Review*, Mar–Apr.

Franks, J. (2000). Supply chain innovation. *Work Study*, 49(4), 152–5.

Gibb, A. G. F. (2000). *Client's Guide and Tool Kit for Standardisation and Pre-assembly*. Construction Industry Research and Information Association (CIRIA) Report CP/75.

Holti, R. *et al.* (1999). *Prime Contractor Handbook of Supply Chain Management*. Defence Estates.

Koskela, L. (1992). *Application of the New Production Philosophy To Construction*. CIFE Technical Report No. 72, Stanford University.

SAVE International (1997). *Value Methodology Standard*. SAVE.

Womack, P. and Jones, D. (1996). *The Machine that Changed the World*. Touchstone Books.

Womack, P. and Jones, D. (1998). *Lean Thinking*. Touchstone Books.

3

Managing value.
Part 2: Partnering

Douglas I. Gordon FRICS

Introduction

This chapter examines the role of partnering in procurement and in supply chain management, which is described as the process that welds the supply chain together, and takes the reader through a step-by-step explanation of the process as well as discussing the opportunities that this approach offers to the quantity surveyor.

Overview – a client's perspective

There is clear evidence that the UK construction industry is following in the footsteps of the offshore oil and gas industries by adopting partnering as one of the measures necessary to cure some of its identified problems. The public and private sectors appear to have accepted Latham's recommendation, where he expressed the hope:

> ... that the industry and its clients will now ... embark upon partnering. The industry can then concentrate on improving its performance and serving its clients better, without wasting time and energy on adversarial conflict.

Following on from this, government ministers have repeatedly expressed the wish to see partnering become the norm in the hope that it will promote a new way of working.

As if to reinforce the message coming from the public sector, in 1998 the Construction Clients Forum identified that only contractors adhering to the language of partnering would qualify for the work let by its members, who are drawn from the public and private sectors and account for the commissioning of 80% of UK projects by value. The Housing Forum, representing public, social and private sector organisations, established a Partnering Working Group in 1999 to produce a model partnering agreement 'which would encourage all organisations in the housebuilding sector to look at partnerships as a way of achieving the targets of the Egan Report'. There is further evidence of the growth of the acceptance of partnering in the National Health Service ProCure 21 programme, the aim of which 'is to introduce long-term relationships between the NHS and the construction industry to establish partnering arrangements and effective supply chain management'. NHS ProCure 21 is intended to promote better capital procurement by establishing a partnering programme, through the development of long-term framework agreements with the private sector, which continuously maximises the benefits of supply chain management. This approach will, it is hoped, provide the opportunity to integrate design and construction and (where appropriate) facilities management in order to exploit the skills of all participants in optimising functionality and buildability, delivering better value for money, and ultimately providing a better service for patients. This 4-year framework programme, which was introduced in 2001 and covers a series of projects, is in its pioneering stage, and represents the NHS Estates' prime contracting initiative for procuring building work below the £20 million PFI threshold. If successful, it is expected to become the standard procurement method across the NHS, eventually expanding to including large PFI projects.

It is clear that public sector clients (including local authorities) are being directed towards procurement strategies that are based on integration and collaboration. The National Audit Office's report, *Modernising Construction*, gave support for public sector clients in promoting innovation and good practice, encouraging the industry as a whole and its clients to:

- Select contractors on the basis of value for money
- Develop close working relationships between clients and the entire supply chain
- Integrate the entire supply chain, including clients, professional advisers, designers, contractors, sub-contractors and suppliers.

In the private sector, many major and influential clients across all sectors have been adopting partnering in response to the proven long-term benefits that can be achieved through this approach. There is, however, little evidence that small, occasional clients have much to gain from the process.

Overview – a contractor's and consultant's perspective

So much for the clients; what about contractors, their supply chains, and consultants? What does partnering hold for them? Main contractors have enthusiastically embraced the partnering concept, without which much of the work available from the major clients is not accessible. General and specialist subcontractors, suppliers and manufacturers may be involved through partnering within a larger supply chain, but many claim that the only benefit for them is assured workload, although this comes at a price – lower profit margins, for example. Nevertheless, it is interesting to note the growth of networking events/marketplaces that offer manufacturers the opportunity to forge new relationships by providing a consultation service on their stands to promote their design and problem-solving skills, rather than selling products. These collaborative, solutions driven events have been enthusiastically supported by major materials manufacturers, who recognise the contribution they are already making to design through early involvement in partnering arrangements. This attitude has resulted in the recent formation of the Construction Manufacturers' Partnering Association (Compass) to facilitate engagement in strategic partnering ventures.

Many opinions have been published regarding partnering between clients and contractors (including specialists); far less appears about partnering with consultants. Many of the large client organisations have framework agreements with consultants as well as contractors, covering periods of time and a series of projects, and of course many consultants have been 'preferred' firms of regular clients for many years. Framework agreements often do little more than formalise long-term relationships, although some clients are becoming more demanding of their consultants in these agreements, resulting in firms being dropped and others refusing to sign up. The main attraction for consultants, large or small, is undoubtedly the security of workload offered by long-term arrangements, but this may be at a price – financial or otherwise. Many consultants, particularly architects, have formed strong relationships with contractors to compete for design and build projects, which have been increasingly attracting many clients for some years. Some of these arrangements are now developing into the core of prime contracting alliances. Besides security of workload, attractions of partnering for consultants might include the satisfaction and reputation gained from being associated with successful projects or high-profile clients. Theoretically, greater profits are achievable through sharing in savings; however, there is a lack of hard evidence that consultants benefit from this. Nevertheless, it is likely that, where consultants partner with contractors in, say, a design and build or prime contracting project, the consultants may well share some of the savings awarded to the contractor. Project specific partnering would not appear to offer many benefits for consultants.

Managing the supply chain

The preferred approach to managing the supply chain is partnering – it welds the links of the supply chain together. Some enthusiasts would claim that partnering is a panacea for the industry's ills; pragmatists recognise that it is only one of many solutions that may be appropriate for some, but not all, situations. The Construction Industry Board's report *Partnering in the Team* (1997) states that:

> It is acknowledged that partnering is not an appropriate procurement strategy for all construction projects or programmes ... Partnering succeeds best ... where ... the project or programme is high value and high risk [and] the contractor's interest is fuelled by the prospect of a high value/high attractiveness account core to their business.

Here, the contractors' interests relate to high turnover and secure profit levels.

Although the term 'partnering' is relatively new, having been adopted in various guises within the UK construction industry since the late 1980s, this is not the case with the relationship itself. Some contractors had been practising what they might term collaborative contracting for many years before the term partnering was adopted with respect to a formal arrangement – for example, Bovis' relationship with Marks and Spencers.

Essentially, partnering is a process that enables organisations to develop collaborative relationships either for one-off projects (project-specific) or as long-term associations (strategic partnering). The process is used as a tool to improve performance, and may apply to two organisations (e.g. a client and a design and build contractor) or to a number of organisations within a formal or informal alliance (e.g. consultants, contractors, subcontractors, suppliers, manufacturers, etc., with or without client participation). The partnering process is formalised within a relationship that might be defined within a charter or a contractual agreement.

The National Economic Development Office report *Partnering: Contracting without Conflict* (NEDO, 1991) adopted the following definition, which was originally produced by the United States Construction Industry Institute (CII):

> Partnering is a long-term commitment between two or more organisations for the purpose of achieving specific business objectives by maximising the effectiveness of each participant's resources. The relationship is based upon trust, dedication to common goals and an understanding of each other's individual expectations and values. Expected benefits include improved efficiency and cost effectiveness, increased opportunity for innovation, and the continuous improvement of quality products and services.

The NEDO report interpreted this as meaning that:

> ... partnering is a contractual arrangement between a client and his chosen contractor which has a term of a given number of years rather than the duration of a specific project. Thus ... contractors may be responsible for a number of projects...

Clearly this relates to strategic partnering. Bennet and Jayes (1995) streamlined the definition, and identified that it can be project specific. They also added 'method of problem resolution' to the process, and 'measurable' to continuous improvement. Their definition also makes reference to it being a 'management approach' rather than NEDO's reference to a contractual arrangement. In the report, *Trusting the Team*, partnering is defined as follows:

> Partnering is a management approach used by two or more organisations to achieve specific business objectives by maximising the effectiveness of each participant's resources. The approach is based on mutual objectives, an agreed method of problem resolution and an active search for continuous measurable improvements. Partnering can be based on a single project (project partnering) but greater benefits are available when it is based on a long-term commitment (strategic partnering).

This reflects how the perception of partnering had changed in a relatively short period of time. The understanding of the process had further developed by 1997, when Bennet and Jayes, with the Partnering Task Force of Reading Construction Forum, published *The Seven Pillars of Partnering*. This report defined partnering as:

> ... a set of strategic actions which embody the mutual objectives of a number of firms achieved by co-operative decision-making aimed at using feedback to continuously improve their joint performance.

Here, problem solution is subsumed within decision-making in what they term 'second generation partnering'. This approach, which recognises that more significant benefits can be achieved through long-term relationships throughout the supply chain, 'begins with a strategic decision to co-operate by a client and a group of consultants, contractors and specialists engaged in an ongoing series of projects'. The Forum's research suggests that a third generation is emerging where organisations form an alliance to utilise fully the expertise within the supply chain in order to provide comprehensive packages (similar to the PFI) where clients wish to outsource construction (and perhaps facilities management) and concentrate on their core business.

Partnering is a structured management approach to facilitate teamwork across contractual boundaries that helps people to work together effectively in order to satisfy their organisations' (and perhaps their own) objectives. It is seen by many as a means of avoiding risks and conflict. There isn't one model partnering arrangement; it is an approach that is essentially flexible, and needs to be tailored to suit specific circumstances.

Words and phrases selected from numerous definitions or descriptions of partnering relate either to the process (such as: avoid waste and conflict; effectively communicate; co-operate/ collaborate; work together/teamworking; integrated; co-ordinated; continuous evaluation; avoid/minimise/share risk) or to the desired outcomes (such as: achieve mutual/common objectives; achieve continuous measurable improvement; maximise/share benefits/rewards).

The pros and cons of partnering

Mutual objectives and areas for continuous improvement are listed below, and indicate the benefits that are claimed to be achievable from partnering:

1. *Improved quality*:
   - Increased quality through consistency
   - Increased reliability
   - Reduction in defects/faults = less rework/zero defects
   - Facilitates total quality management.
2. *Improved design*:
   - Early input of supply chain offers more effective solutions
   - Increased buildability through integration of supply chain
   - Reliable information
   - Faster flow of information
   - Improved product development.
3. *Improved production* (particularly with long-term relationships):
   - Increased efficiency and avoidance of waste
   - Reduction in learning curves due to continuity/standardisation = increased productivity
   - Long-term planning = investment in technology, recruitment and training = greater efficiency
   - Focus on continuous improvement = improved process
   - Reduction in restrictive practices/development of multi-skilling.
4. *Improved time*:
   - Increased speed
   - More certainty of timescale/completion dates.

5. *Reduced costs*:
   * Reduction in capital cost and more certainty of final cost
   * Lower legal costs due to open communication and avoidance of conflict
   * Encourages innovation, leading to less cost/extra value
   * Lower whole-life costs due to greater interaction between client and supply chain
   * Reduction in duplication of effort.
     With long-term relationships:
   * Reduction in marketing/tendering costs
   * Savings in mobilisation/demobilisation
   * More responsibility devolved to contractors, who can optimise their organisation.
6. *Better risk management*:
   * Collective risk management
   * Problem-solving/win–win attitudes, which lead to avoidance/reduction of risks
   * Shared risk and responsibilities/absence of blame culture
   * Client does not pay for risks that do not materialise.
7. *Increased value*:
   * Increased value and fitness for purpose through better supply chain management with a focus on client satisfaction
   * Value engineering encouraged
   * Increased value due to improvements in quality, time, cost and risk
   * Sharing of information increases value to all
   * 'Two heads are better than one' = better decisions
   * Release of staff from conflict issues creates more time for adding value
   * Improved product due to focus on continuous improvement
   * Concentration on core activity/business = focus on what people/organisations are good at.
8. *Increased profits*:
   * Potential to be higher
   * Related to added value
   * Sharing of savings
   * More assured and greater stability of business.
9. *Improved cash flow*:
   * Through better relationships
   * Often no retention held.

10. *Improved administration*:
   - Reduced overheads
   - Less paperwork
   - Less policing of contractors
   - Fewer and more efficient interfaces
   - Simpler project organisation
   - Eliminates duplication of resources
   - Effective communication.
11. *Improved safety*:
   - Through integration of design and construction
   - Through consistency.
12. *Contractor's workload* (with long-term relationships):
   - More certainty of volume
   - Continuity of work and use of resources.
13. *General issues*:
   - Enhanced reputation through association with successful projects
   - Greater trust, communication and consultation
   - Greater understanding of each other's objectives
   - Mutual respect and support
   - Better staff development
   - Development of 'can-do' attitude/problem-solving skills
   - Better rewards/career opportunities
   - Avoids conflict (through collaborative teamworking, open communication, better understanding and agreed problem resolution procedures), which leads to less confrontation, better working conditions, enhanced employee satisfaction/morale/team spirit, greater job interest, greater motivation and commitment, lower staff turnover, greater efficiency, and ultimately greater value
   - Leads to reduction in the industry's fragmentation
   - Greater dissemination of knowledge, skills and expertise, resulting in an improved industry.

Some of the above will benefit the client and others will benefit the contractor; collectively, they should lead to satisfaction for both. The extent and nature of the specific arrangements in place will determine whether or not others in the supply chain (e.g. consultants, subcontractors, suppliers, manufacturers) will profit from partnering. Where contractors faithfully adopt the partnering philosophy within their own supply chain management, the benefits for those involved are obvious – they will share in the contractors' own benefits. However, there are claims that many contractors do not share their rewards, and indeed that some main contractors simply pass the costs/risks down the supply chain. This is discussed later.

If these claimed benefits are achievable, why is partnering not universally adopted? There is a significant amount of literature expounding the benefits, some claiming large savings in cost and time; however, although many case studies are cited, as Fisher and Green (2001) indicate, 'Independently verified evidence of measurable benefits is currently lacking in sufficient quantity and breadth to be convincing'. Also, as stated earlier, partnering is not the answer for all circumstances, and of course, there can be a downside. In addition, its adoption is not necessarily easy.

Considering the downside, some of the potential problems stem from cultural issues, some are caused by the fact that partnering exists alongside the conventional procurement strategies, and others are caused by the partnering process itself. Issues that need to be considered include the following:

- Successful partnering is dependent upon the people involved. Until partnering has become an accepted way of doing business for a significant period of time, many of these people will have been educated, trained and experienced in an industry that is unquestionably riddled with adversarial attitudes. A culture change and development of new attitudes is necessary; but this is difficult, even with re-education. Trust is an essential factor in partnering, and is not in great abundance within today's construction industry. At present far too many organisations are jumping on the bandwagon or are being forced into participating in the process, without the personnel necessary to ensure success.
- The alignment of attitudes and objectives can be difficult between culturally different organisations; indeed, difficulties are often apparent internally within large organisations.
- Difficulties are often experienced with the concept of mutual objectives when using standard forms of contract, such as JCT, ICE, etc., which could be described as rules of combat.
- Where a charter is used, it is often not compatible with the contract.
- Partnering can involve the commitment of resources upfront, which, for some partners on large projects/programmes of

work, will not reap rewards for a considerable period of time. This becomes more acceptable in strategic partnering, where the benefits, once they arrive, should flow on a regular basis.

- There can be difficulties in equitably sharing commitment, risks and rewards.
- Many small organisations find that their expected commitment can mean excessive time being spent at meetings and undertaking other forms of communication.
- It is claimed that consultants are sometimes involved in extra work due to the increase in alternative solutions that arise and needing consideration. This can result in professionals needing to be paid more for their input, or they need to undertake more work for their share of the rewards.
- There is the possible loss of confidentiality with respect to proprietary information and resultant loss of competitive edge.
- There is the risk of increasing interdependency amongst partners in strategic relationships, and ideas becoming stale due to lack of new players.
- There is the danger that players may become complacent and relationships become too cosy. Clients can lose touch with the market, and the benefits of downward trends may not be passed on to clients tied into long-term agreements.
- Highly integrated supply chains can lead to lack of competition.
- Powerful partners often take advantage of their situation and grab the lion's share of rewards and/or pass costs and risks down the supply chain. This can be particularly noticeable where clients require performance targets to be progressively adjusted for continuous improvement, which may result in lower margins for contractors, resulting in their supply chain being squeezed. Many of the smaller players may accept this to maintain a relationship and secure turnover.
- Some clients consider that they may be given contractors' or consultants' B teams, with A teams being reserved for competitive projects. However, some others take the opposite point of view, which can result in them not considering supply organisations who already have strong relationships with other clients, assuming that the A teams are already committed. This latter situation can be relevant where clients in a particular sector are few – a situation that has

been evident in the UK offshore oil and gas industry.
- Some organisations are reluctant to enter into an open-book culture.

Provided that participants in the process are aware of the potential downside issues, measures can and should be taken to counteract the undesirable effects. These are included later in the prerequisites/key success factors.

In order to achieve benefits from partnering, it is now almost universally agreed that the best results are achieved through long-term agreements and partnering the entire supply chain:

> The benefits of partnering relationships are cumulative, so that strategic alliances produce significantly more advantage than single project arrangements. And the benefits are significantly greater if partnering is applied throughout the supply chain.
>
> (Construction Industry Training Board, 1997)

> An essential ingredient in the delivery of radical performance improvements in other industries has been the creation of long-term relationships or alliances throughout the supply chain on the basis of mutual interest. Alliances offer the co-operation and continuity needed to enable the team to learn and take a stake in improving the product. A team that does not stay together has no learning capability and no chance of making the incremental improvements that improve efficiency over the long term. The concept of the alliance is therefore fundamental to our view of how efficiency and quality in construction can be improved and made available to all clients...
>
> (*DETR*, 1998)

This has been accepted by the public sector:

> The greatest benefits from strategic partnering arrangements arise because the lessons learnt from one project can be applied to further similar projects through a process of continuous improvement. Strategic partnering arrangements should be adopted in preference to project specific partnering arrangements wherever possible. Irrespective of the type of

partnering relationship that the client enters into with a primary supplier (such as a main contractor or main consultant), significant benefits (in achieving overall value for money) can be obtained where a primary supplier has entered into strategic partnering arrangements with secondary suppliers (such as sub-contractors or sub-consultants). Supply chain relationships of this type are essential to obtain the maximum benefits from partnering.

(HM Treasury, 1999)

The European Community competition regulations regarding public sector procurement are seen by some as problematic when considering strategic partnering; however, these regulations are not necessarily incompatible with the public sector best value policy. European Union regulations permit partnership agreements and selection of contractors based on the following criteria:

- Previous experience with client
- Capability and track record
- Health and safety record
- Quality systems
- Financial data
- Overheads and profit margins
- Experience of partnering the supply chain
- Management of the supply chain
- Experience of process improvement tools such as value management, value engineering, risk management, and continuous improvement.

Under the EU regulations, framework agreements may be entered into covering a programme of work or series of projects (see Chapter 7).

Key success factors

Simply adopting a policy of partnering with, and within, the supply chain will not itself ensure success. Partnering is not easy; a number of prerequisites, or key success factors, need to be taken on board. Some of the following are desirable for project partnering; all are essential for successful strategic partnering:

- The decision to enter into any partnering agreement, particularly where it means long-term commitment, needs to be based on sound reasoning and an awareness of the implications. There needs to be a commitment at all levels within an organisation to make the project or programme of work a success, which means a commitment to working together with others to ensure a successful outcome for all participants (win–win situation). There must be an understanding that this will only be achieved through sustained improvement in quality and efficiency. Returns will not be immediate; a willingness to make an early (and perhaps not insignificant) investment in time and effort to build the team is essential.
- Once a decision has been made, it is vitally important to ensure that a careful choice is made with respect to partners. Partners must have confidence in each other's organisations, and each organisation needs to have confidence in its own team, which means careful selection of the people involved. Participants need to have a clear understanding and commitment to the teamworking culture. Partners should be chosen on the basis of the ability to offer best value for money and not on lowest price; their ability to innovate and offer effective solutions should also be considered:

  > Partnering is only appropriate between organisations where top management share the fundamental belief that people are honest, want to do things which are valued, and are motivated by challenge. Such organisations trust their people and seek ways to enable them to add value to their business.
  >
  > (Construction Industry Training Board, 1997)

  Clients should normally select their partners from competitive bids based on carefully set criteria aimed at getting best value for money. This initial competition should have an open and known prequalification system for bidders.

- Partners need collectively to agree the objectives of the arrangement/project/programme of work and ensure alignment/compatibility of goals. This will require early involvement of the entire team to ensure a win–win situation for all. It is this agreement that should drive the relationship, not the contract(s). The agenda must be mutual interest with a focus on the customer; it must therefore be quality/value driven.

- To satisfy the relationship's agenda, there needs to be clarity from the client and continued client involvement. It is essential to define clearly the responsibilities of all participants within an integrated process. There can be no weak links. People, without regard to affiliation, must be brought together into integrated teams with streamlined supply chain management. There needs to be a willingness to be flexible and adopt new ideas and different ways of doing things – e.g. different operational methodologies, different administrative procedures, different payment methods, different payment procedures, etc.

- Sharing is important. All players should share in success in line with their contribution to the value added process (which will often be difficult to assess). There also needs to be a sharing of information, which requires open-book accounting and open, flexible communication between organisations/teams/people. Responsibility for risks must be allocated clearly and fairly, but there must be a collective responsibility for problems and an openness and willingness to accept and share mistakes. This requires a departure from the finger-pointing, blame culture to an acceptance that getting things wrong results in a lose–lose, rather than a win–win, scenario. Adoption of such openness and sharing requires trust.

- It is important that all partnering arrangements incorporate effective methods of measuring performance. It has been identified that partnering should strive for continuous improvement, and this must be measurable to ascertain whether or not the process is effective. It is essential therefore that agreed measurable (and achievable) targets for productivity improvement are set, and clear, easily understood measurement systems are adopted to evaluate efficiency with respect to time and cost. Similarly, ways of measuring improvements in quality must be adopted as part of the quality assurance procedures. Benchmarking, perhaps including use of the industry's key performance indicators (KPIs), should be adopted and, to enable organisations to share in savings/increased benefits, effective incentive schemes need to be in place. In long-term agreements, annual reviews should take place to reset objectives/targets. Clients with major programmes of work should not put all their eggs in one basket; this will enable them to benchmark their different partners against each other. These clients may periodically test the market by limiting the time of agreements to enable new bids to be made; however, it is important that the duration of agreements is of reasonable length to assure bidders of predictable turnover (subject to satisfactory performance). This will, of course, require clients to be able to assure continuity (or at least predictability) of workload.

- There will be times when partners don't agree, and it is therefore important that agreed non-adversarial conflict resolution procedures are in place to resolve problems within the relationship. The principle of trying to resolve disputes at the lowest possible level should normally be adopted to save time and cost.

- As identified earlier, a culture change is needed for individuals (and organisations) used to the conventional ways of the industry. Education and training is needed to ensure an understanding of partnering philosophy. However, it is important that, regardless of how well versed participants are in the philosophy and procedures, teambuilding takes place at commencement of the relationship. Good facilitators and change management and behavioural management specialists should be used for personal and team coaching. The 'us and them' barriers need to be broken down; people need to be encouraged to listen to and learn from others so that the attitudes of 'contractors don't understand design' and the 'architects don't understand construction' can be eradicated; there has to be an acceptance that no party benefits at the expense (or from the exploitation) of another; and good interpersonal relationships need to be developed at interfaces. All of this requires good teambuilding.

- Good teambuilding and development of efficient teamworking can be enhanced by adopting commonsense procedures such as sharing of office space and sharing of information through computing networks/intranets, etc. Successful teambuilding should result in the development of trust.

Trust is generally regarded as being crucial to the success of partnering; indeed it has been described as the cornerstone of a successful partnering relationship. Trust must exist within organisations – i.e. individuals need to be trusted – and between organisations – i.e. organisations need to trust each other. However, Blois (1999) argues that only individuals can trust, and consequently trust between organisations (which are, after all, only collections of people) means a lot of people needing to trust a lot of other people. This makes relationships vulnerable to human fickleness. Careful selection of the correct individuals by each organisation is therefore crucial to success. These individuals, particularly those at the interface with partners, must fully accept the philosophy of partnering and be committed to the success of the arrangement. Changes in personnel or attitudes of key people can render relationships fragile; it is therefore very important to invest in trust-building activities amongst everyone involved with integrated teams so that the collective trust in the team as a whole can withstand breakdown in individual relationships. The building of trust, which is essential for co-operative activity, takes time, effort and patience, particularly for those with experience of the traditional way of doing business. Where trust has been established, partners can be relied upon. However, Blois states that:

> ... while trusting someone implies being willing to rely on them, the opposite is not necessarily the case. What distinguishes trust from reliance is the expectation that the other party may take initiatives (or exercise discretion) to utilise new opportunities to our advantage, over and above what was either explicitly or implicitly promised.

Reliance on others to do what is expected is necessary in any arrangement, including conventional procurement strategies, where contract is used to encourage/enforce it. This extra ingredient, which stems from trust, is what is being sought within partnering arrangements. There will be times when organisations in a partnering arrangement are tempted not to act in a mutually acceptable way. This may be for short-term gain, or market changes may result in changes in attitude. In such times, many organisations may be tempted to break faith and break ranks. Only when trust has been fully developed will commitment to seeing things through be expected of partners.

The partnering process

The process involved in partnering will vary considerably depending on the type of partnering – project or strategic – and the organisations involved. Partnering may, for example, be throughout a supply chain, but not include the client. Where a client is involved, the arrangement could be:

- Between client and main contractor only
- Between client and the leader of a supply chain (e.g. a design and build contractor or prime contractor)
- Between client and various organisations (such as consultants, constructors and suppliers), either as separate individual relationships or one collective relationship.

Of these, the first is not uncommon; it is straightforward but clearly limited in effectiveness. The second option is far more effective, especially where partnering also takes place within the supply chain, and is quite straightforward and easy to understand. Current trends indicate that this is becoming a favoured option. The final option gives the client more choice than the second option – e.g. major clients can bring together different organisations with whom they partner to suit different projects. This is, however, less likely to produce the same quality of teamwork as the second option.

Typically, the following process would normally be adopted:

1. *Decision to partner.* The decision needs to be made on rational consideration of the potential benefits and the required commitment, with a clear focus on why the process should be adopted.
2. *Choice of partner(s).* Careful selection of partner(s) is essential, particularly in the case of long-term arrangements. Decisions may be based on existing relationships or past experience, but will at times (particularly in the public sector) be based on initial competition as discussed earlier.
3. *Initial workshop.* A workshop should take place at the commencement of the relationship to clarify the rationale, set goals and objectives, and agree the ground rules. This workshop would normally last from 1 to 3 days, be held in a neutral location and be attended by all key personnel from the partners. The workshop should be run by a neutral facilitator

with experience of partnering. Detailed analysis of all parties' strengths and weaknesses should be undertaken and key resource contributions identified. Mutual agreement to the objectives, relating to such aspects as quality, time, costs, safety, etc., should be established, together with a plan as to how they can be achieved. Responsibilities should be identified and a conflict resolution mechanism agreed to. Any incentive mechanisms should also be agreed upon, targets should be set and methods agreed for measuring continuous improvement. Leadership of the team will be important. Often it is either a client representative or the main contractor who leads; however, any key player could be chosen. An essential factor is choosing a person with good experience of the process. This is the time for commencing the team-building and trust-building processes. Individual and collective commitment needs to be made at this time, and decisions need to be made not only on the details of the relationship but also on its format – e.g. will a charter, a contract or both be used to seal the agreement?

4. *Implementation.* Once agreement has been reached, the relationship can be implemented and regular workshops should be undertaken to reinforce the culture, continue the team building and trust-building processes, and monitor and review progress.

5. *Completion.* On completion of a project-specific agreement, or perhaps annually on strategic agreements, there needs to be analysis of performance and feedback. Is it worth doing again/continuing?

Previously measurable continuous improvement and incentive schemes have been mentioned. The use of the industry's KPIs (see Chapter 1) or similar seems to be a favoured way of measuring the performance of the relationship. It is likely that the achievement of many objectives can be ascertained using these types of indicators, specifically or collectively. The results should indicate the success, or otherwise, from the client's point of view. The supply chain as a whole will be able to identify its performance, and particular indicators may provide meaningful feedback to individual organisations; however, success for many of these organisations will be measured in commercial terms. Where incentive schemes are used, for example to share savings, the extent of the rewards will indicate the success or otherwise for many players.

It is important that incentive schemes should lead to added value for the client and increased rewards for the supply team. Rewards should not be given for improved performance that does not benefit the client; it is important, therefore, that only performance targets meaningful to the objectives agreed at the outset should be used. When choosing an incentive scheme, it is important to bear in mind that the spirit of collaboration in partnering should be reflected in the sharing of risks as well as rewards. The mechanism chosen should therefore penalise poor performance as well as reward exceptionally good performance.

The consensus of opinion is that open-book accounting provides the best basis for measuring performance. The target cost method is frequently adopted. This is a clear, easily understood way of calculating savings or overruns in cost. Typically, cost targets are set for activities, operations, packages, projects or periods of time, and contractors are reimbursed their costs (usually production costs plus direct overheads) plus a variable margin (to cover indirect overheads and profit). Various methods can be used to calculate the margin payable, e.g. an agreed amount that can be adjusted by:

- Sharing cost savings or cost overruns in agreed proportions
- Payment of bonuses, or deduction of penalties, which kick-in at specific cost milestones
- Performance measures, perhaps including time, defects/quality, safety, etc., as well as productivity.

Various mechanisms that reflect the range and relative importance of the objectives set for the particular partnering agreement can and should be used. These examples are relatively straightforward when considering risk and reward shared by, say, a client and a design and build contractor (or even including that contractor's supply chain of subcontractors and suppliers); however, it becomes more complex and difficult equitably to share amongst an entire supply chain including consultants. In such cases it may be difficult to ascertain which organisations contributed, and to what extent the contributions affected performance. Within alliances, such issues are not so significant; sharing is likely to be freer, and possibly based on the proportion of partners' inputs, with swings and roundabouts ensuring equity over time for contributing organisations (those not contributing will be replaced).

The initial workshop will normally culminate in the production of a partnering charter. This will often comprise a mission statement and a list of objectives that will express how the participants intend to act and what they wish to achieve. Content will vary, but will normally address aspects such as relationships (trust, co-operation, etc.), quality, time, cost, value, profit, cash flow, sharing of savings, safety, and reputation. In addition, agreed methods of implementing the arrangement need to be set down – e.g. methods of measuring performance, review procedures, risk allocation, dispute resolution procedures, etc. There are two schools of thought on how this should be done. One view is that the participants should act in good faith and only sign a non-binding partnership charter, while the other view is that there must be a contract to define clearly aspects such as participants' rights and obligations, time for performance, allocation of risk, consequences of failures, etc. Much has been written on the debate and no doubt will continue to be, and no attempt is made here to draw conclusions from the debate or to suggest which course of action should be taken. A range of options is now available, and these are described below.

The Construction Task Force stated in *Rethinking Construction* that it 'wishes to see an end to reliance on contracts … If the relationship between a constructor and employer is soundly based and the parties recognise their mutual interdependence, then formal contract documents should gradually become obsolete'. It suggests that if it can be done by the motor industry (quoting the non-contractually based relationships between Nissan and its principal suppliers), then the construction industry should be capable of doing it. Construction projects with no formal written contract have taken place with apparent success; however, most participants appear to favour the use of some form of contractual agreement. *Partnering in the Team* (Construction Industry Training Board, 1997) states that partnering 'does not remove the need to use a recognised modern form of contract… It is, of course, expected that the spirit in which the works are carried out under a partnering agreement will minimise the need to resort to the contract'.

Those favouring the no-formal-contract approach consider that the use of contracts is alien to the ethos of partnering and, where standard forms like JCT and ICE are used, the development of harmonious relationships is stifled. They rely on the content of the non-binding partnering charter to achieve the team's objectives. Those wishing to have a contract have choices.

A bespoke legally binding agreement can be created, which backs up (or incorporates) the partnering charter. This of course carries the normal dangers associated with non-standard contracts. Where incorporated into one document, there is a danger of confusing the charter and the contract. The purpose of the charter is to express the commitment to work collabora- tively to achieve win–win solutions; it should not fundamentally alter the nature of the parties' legal and contractual relation- ships. Rights and obligations are dealt with by the contract.

The use of non-binding partnering charters backed up by standard forms is common. Often the standard forms are amended, e.g. removal of liquidated damages and retention clauses. Those using standard forms often 'put them in the drawer' as a fallback only in the event of something going seriously wrong. The JCT Practice Note 4 provides a 6-page summary of the principles of partnering, and includes a non- binding partnership charter for single projects that can be used with existing forms of building contract. This approach is suitable for those who wish to work within a framework with which they are familiar.

Where a New Engineering Contract (NEC) such as the Engineering and Construction Contract (ECC) is used, a partnering agreement is in principle achieved between the two parties. The aim of the ECC is to provide a modern method for employers, designers, contractors and project managers to work collaboratively. ECC Clause 10.1 requires parties to act 'in a spirit of mutual trust and co-operation'. The NEC states that 'A partnering contract, between two parties only, is achieved by using a standard NEC contract' (from the guidance notes to the NEC Partnering Option). ECC therefore creates a legally binding agreement (a partnering contract) between a client and a contractor. Similarly, the Engineering and Construction Subcontract and the Professional Services Contracts create agreements between the relevant parties. A partnering option exists for use where more than two parties are partnering on the same project or programme. This option is given legal effect by including it in each relevant bi-party contract by means of an additional clause (Option X12 in the ECC, for example). This does not create a multi-party contract, but will tie in all partner- ing team members, through their bi-party contracts, to the

objectives of the partnering team. This option imposes responsibilities additional to those in the basic NEC that relate to the
partnering objectives. The partnering process is managed by a
core group, which acts and takes decisions on behalf of the
partners. This core group is normally led by the client's representative and will attempt to resolve problems between partners
who do not have a contract between themselves.

The Association of Consultant Architects (ACA) has published
the first standard form Project Partnering Contract (PPC 2000)
intended for multi-party partnering. This allows the client, the
contractor, consultants and key specialists to sign a single
partnering contract. Like the NEC, provision is made for a core
group of key individuals representing the team members. This
contract also introduces a new role – that of partnering advisor,
described in the explanatory notes as 'an individual with
relevant experience who can guide the partnering process, who
can document the relationship, commitments and expectations
of Partnering Team members and who can provide an additional
facility for problem resolution'. Clearly this is a key role in a
new contract, which is expected to accelerate culture changes
and improve supply chain management and performance.

Whatever contract is chosen, problems may lie ahead with
enforcement of 'good faith' obligations under English law. This
need not be a problem with arbitration, but where litigation is
resorted to, unless the courts change their attitudes to such
obligations (which are, of course, central to partnering philosophy) the intention of parties may not be enforceable unless obligations are explicitly expressed. There is little doubt, however,
that the NEC contracts (with the partnering option) or PPC
2000 will be chosen by many clients. These contracts, particularly PPC 2000, make partnering with the entire supply chain,
including consultants, more straightforward.

The future of partnering

There seems to be little doubt that the industry is accepting the
need for collaborative working. This would appear to be best
achieved by formalising relationships in a partnering arrangement. Trends have for some time shown an increase in the use
of design and build procurement in the private sector, mainly
based on the desire for single-point responsibility and reduction

in client risk. Following Latham and Egan, the public sector has introduced a policy that virtually insists upon the use of procurement methods that integrate design and construction in the pursuit of best value, and also makes strong recommendations regarding the use of partnering. There is general consensus that long-term relationships provide the best opportunity for greater efficiency and increased value. The result of this is a drive towards clients with regular needs for construction and maintenance of property opting for long-term partnering arrangements, either individually with chosen contractors and consultants or collectively with an entire supply chain. There is little doubt in most people's minds that the most efficient supply chains will practise partnering within the organisation, and the smart client will partner with them. Modern standard forms of contract are facilitating this process.

The evidence available indicates that there are benefits for regular clients and for main contractors. Specialist contractors can also benefit, particularly when clients recognise the need to involve them early. Where good supply chain management exists, there are also benefits for subcontractors, suppliers and manufacturers. Medium and small contractors should not feel that they are excluded. The large main contractors tend to hit the headlines, but there are regular clients with small projects and programmes that are not suitable for the larger organisations.

Partnering – blessing or threat?

The future for consultants is less clear-cut. The skills of the professional continue to be in great demand; in general, most individuals do not face a threat from partnering. Consultant practices do face threats, but there are also exciting opportunities for those firms willing and able to grasp them. There is little doubt that some consultants are feeling marginalised by the development of close relationships between clients and contractors, particularly where contractors can offer an integrated supply chain including design. Doors are being closed to architects by, for example, the public sector policy of choosing integrated design and construction. Unless these architects join some form of alliance with contractors or construction managers, this work is not available to them. This problem for designers does not apply equally to quantity surveyors, many of whom have already found alternative roles in design and build projects, advising clients or supporting contractors. There is however little doubt that a move away from the separate design and construction processes with competitive tendering to select contractors leads to a decrease in the traditional quantity surveying activity.

The design consultants (e.g. architect and engineers) need to make choices. They can work in traditional non-integrated ways for clients who don't want to partner and/or don't want a design and build type project. Outside the public sector, there will be many such clients. Where they are willing to partner with clients they should market themselves accordingly, and enter into framework agreements or more loosely structured long-term relationships. Where they are willing to collaborate with constructors, they will need to consider developing a close working relationship with one or more contractors to secure design and build projects, or to form or join an alliance with a group of organisations which creates an integrated supply chain capable of bidding for prime contracts. Those wishing to operate outside the integrated systems may nevertheless find opportunities as independent advisors to clients choosing such systems (e.g. conceptual design).

Members of the Construction Task Force have been quoted as saying that the quantity surveying profession will become obsolete once the industry becomes less contractual. Such beliefs are clearly based on the adversarial image of the quantity surveyor apparently held by many people in the industry. Whether this is a fair assessment of the quantity surveyor is a matter for conjecture. Some quantity surveyors are undoubtedly adversarial in their approach, but is this inherent in the profession, or have quantity surveyors simply found themselves to be in the best position to provide the industry, or some of its players, with what it wants? To believe that the profession will become obsolete presumes three things:

1. That all (or most) construction projects, will be done in a collaborative way (without a contract?). This is extremely unlikely. The vast majority of clients do not have rolling programmes of construction work; they are occasional or one-off clients only, and will want a contract. Many will want the protection of a quantity surveyor.

2. That quantity surveyors are only employed to deal with contractual issues. This is not the case; they provide a wide range of services.
3. That the quantity surveying profession is not adaptable. However, there is little doubt that the profession has been more flexible than most in adapting to change in the industry.

Collaborative integrated procurement offers opportunities for quantity surveyors, including:

• Acting as an independent client advisor. Many clients will still look to their quantity surveyor for independent advice. This raises the question, 'where is the trust within the relationship if external advice is still needed?' Many clients will still feel that they need advice of someone without an axe to grind – for example, appointing an external quantity surveyor and audit team to ensure that its strategic partners perform. Services provided might include assessment of target costs, development of incentive schemes, measurement of performance, auditing, etc.

• Participating as a partner in an alliance. Quantity surveyors able to demonstrate that they have the skills and ingenuity to add value will be welcomed by most alliances. Imaginative teams will consider numerous solutions – who better to evaluate the alternatives than the quantity surveyor?

• Leading an integrated supply chain. Many quantity surveyors have become successful project managers. There is no reason therefore that they cannot manage a supply chain. With appropriate financial resources, a quantity surveying practice can act as a prime contractor.

• Acting as a partnering advisor within PPC 2000 contracts. The described role would seem to fit the quantity surveyor with partnering experience. This is a key role, and would suit a quantity surveyor who can demonstrate a collaborative rather than adversarial attitude.

The move towards clients partnering with integrated supply chains offers significant opportunities for consultants wishing to join alliances to share in the potential rewards. If the industry does become less adversarial, as is hoped, quantity surveyors will welcome it. They will then be able to concentrate on what they do best – adding value for clients, which coincides with the purpose of partnering!

References

Bennett, J. and Jayes, S. (1995). *Trusting the Team – The Best Practice Guide to Partnering in Construction*. Centre for Strategic Studies in Construction, The University of Reading.

Bennett, J. and Jayes, S. (1997). *The Seven Pillars of Partnering – A Guide to Second Generation Partnering*. Centre for Strategic Studies in Construction, The University of Reading.

Blois, K. J. (1999). Trust in business to business relationships: an evaluation of its status. *Journal of Management Studies*, 36, 197–215..

Construction Industry Training Board, Working Group 12 (1997). *Partnering in the Team*. Thomas Telford Publishing.

DETR (1998). *Rethinking Construction*. Report of the Construction Task Force to the Deputy Prime Minister, John Prescott, on the scope for improving the quality and efficiency of UK construction, the Department of the Environment, Transport and the Regions.

Fisher, N. and Green, S. (2001). Partnering and the UK construction industry, the first ten years – a review of the literature. In: *Modernising Construction,* Appendix 4. HMSO.

National Economic Development Office (1991). *Partnering: Contracting Without Conflict*. HMSO.

HM Treasury (1999). *Procurement Guidance No. 4: Teamworking, Partnering and Incentives*. Procurement Group.

4

Procurement – doing deals

Introduction

This chapter examines the impact of public private partnerships (PPP) and in particular the private finance initiative (PFI) – a procurement strategy described by the Royal Institution of Chartered Surveyors as one of the most important influences on the future of the quantity surveyor during the next decade. Successful PPPs depend upon many of the skills traditionally offered by quantity surveyors, for example, risk management, procurement advice, through-life costs advice etc. The origins, philosophy and motives behind the introduction of public private partnerships are discussed, as well as procedural and contractual aspects of PPPs, the PFI and the Public Private Partnership Programme (4Ps). This chapter will explore the claims and counter-claims put forward by all sides for PPPs, and will conclude by discussing the opportunities that PPPs/PFIs present for quantity surveyors and the trend in adoption of PPPs by other countries.

Public private partnerships

Background and definition

Public private partnerships and the private finance initiative are terms used to describe the procurement processes by which public sector clients contract for capital intensive services from the private sector. Private sector involvement in the delivery of public services in the UK has developed into a very emotive topic, with an unfortunate tendency to generate more heat than light. For many, the confusion and misconceptions surrounding PPP/PFI begin with the definition of these two terms. The term PFI was launched in the early 1990s and then, several years later, the term PPP emerged and appeared to subsume the PFI. The current accepted understanding of the term is as follows:

> Public private partnership is used to describe a range of practical relationships between the sectors that have varying degrees of formality and differing legal or commercial foundations. In its broadest sense, PPP encompasses voluntary agreements and understandings, service level agreements, outsourcing and the PFI.
>
> (Audit Commission, 2001)

Perhaps the adoption of the generic term public private partnership in 1997 also had something to do with creating a softer image of public and private sectors working together, sharing the risk and rewards, in an attempt to counter the public perception that, to date, the PFI had tended to be synonymous with the private sector raping the public sector services (such as the NHS and education) and taking huge profits for shareholders while imposing onerous working conditions on staff running the services.

A PPP project therefore usually involves the delivery of a traditional public sector service and can encompass a wide range of options, one of which is the PFI. In turn, the PFI is one of several similar approaches in a 'family' of procurement that includes design, construct, manage and finance (DCMF), used in the procurement of prisons, and design, build, finance and operate (DBFO), used in the procurement of roads and schools. One of the key objectives of the PFI is to bring private sector management expertise and the disciplines associated with private ownership and finance into the provision of public services. However, if the PFI is to deliver value for money to the public sector, the higher costs of private sector finance and the level of returns demanded by the private sector investors must be outweighed by lower whole life costs.

In a speech to launch the Institute for Public Policy Research's Commission on PPPs on 5 April 2000, Deputy Prime Minister John Prescott declared that: 'Part of the reason for the success of PPP is that there is no one single formula ... It's horses for courses'. The range of PPP models being used in the provision of asset-based services in the UK include, as has already been noted, PFI deals, and these are described in some detail in the following text. There are, however, a number of other potential forms of PPP, including profit-sharing arrangements and long-term contracting, or projects in which the public sector provides the finance and the private sector designs, builds and operates the asset. Examples of existing PPPs are the Channel Tunnel Rail Link, which is backed by a government-backed bond; the London Underground, which involves a private consortium taking over responsibility for investment in and maintenance of the infrastructure assets under a 25–30 year contract while the public sector still remains in control of the operation of services; and the National Air Traffic Control System, where just under 50% the equity in NATS is to be sold to a partner while the workforce together with the public sector retain the remaining majority of the shares.

The development of the private finance initiative

The origins of the UK private finance initiative lie in the introduction, in 1981, of the Ryrie Rules, named after Sir William Ryrie, a former Second Permanent Secretary of the Treasury. These rules were ostensibly a means of allowing private financing, and were developed to try to minimise the impact of government funding restrictions on possibly profitable investment by the nationalised industries. The rules were constantly criticised for being restrictive and giving public bodies little incentive to seek private finance alternatives, and consequently there was almost no use of private finance in infrastructure projects until the construction of the Channel Tunnel in 1987. In an attempt to stimulate increased awareness, in 1989 the Department of Transport published a Green Paper on privately financed roads, *New Roads by New Means*. The Ryrie Rules were partially phased out in 1989, and were finally abandoned in 1992 with the launch of the PFI.

The private finance initiative is the name given to the policies announced by the Chancellor of the Exchequer, Norman Lamont, in the Autumn Statement of 1992. Chancellor Kenneth Clarke's Autumn statements of 1993 and 1994 were used to reshape the design and nature of the initiative. The intention was to bring the private sector into the provision of services and infrastructure, which had formerly been regarded as primarily a public sector concern. For many political spectators the PFI was a natural progression for the Thatcher Government, which had so vigorously pursued a policy of privatisation during the 1980s. In the UK, from the end of the Second World War to 1980 autonomous agencies with political supervision (but not political control) had been responsible for the delivery of electricity, gas, water, telecommunications etc. These utilities were publicly owned assets, and in theory any dividends went to the government. In practice, many of the industries were failing and consequently heavily subsidised by the Treasury and taxpayer in order to maintain employment. Privatisation, it was claimed, would introduce the management skills of the private sector to industries with an ethos of jobs for life and low accountability. In addition, privatisation would open up monopoly markets to other players, with consequent efficiency gains and the bonus of added value for the customer. The process was achieved by valuing the publicly owned assets and then selling them off to the public by means of a share issue open to everyone. The privatisation programme proved to be extremely popular with both the public and politicians, and share issues were heavily over subscribed – the term 'Stakeholder Britain' was coined to described the new phenomenon of widespread share ownership.

However, behind the queues of people waiting at stockbrokers' offices to deliver last-minute applications for allocation of shares there was the less publicised political agenda that continues to make privatisation so appealing to politicians and has led to its spread around the globe like a rash. At the time of the privatisation programme, the Public Sector Borrowing Requirement (PSBR) was the benchmark of a government's ability to control public expenditure. This represented the amount of money needed to be borrowed by government to fund capital projects – the lower the PSBR, the more prudent the government; a point not missed by the electorate, whose taxes in the main funded the capital works programmes. Privatisation is a very easy way for a government to raise revenue, as the whole of the proceeds of the sale of the once publicly owned asset can be set against government debts, reducing budget deficits in the short to medium term and thereby lowering the PSBR and taxation levels. It has been estimated that the British government raised £60 billion by selling previously public owned assets, or what some critics of the programme regarded as the UK's family silver, during the 1980s and early 1990s. Opponents of privatisation also claimed that the money raised could have been substantially higher, as nearly all the public utilities were undervalued by as much as 50% in the government's dash for cash. Not surprisingly, therefore, the PFI has been seen by some as a means of backdoor privatisation of public services, and trades unions, in particular UNISON, have voiced their concerns over the adoption of the PFI. However, as far as government is concerned there is a clear distinction between the sale of existing public assets, which they see as privatisation, and the PFI, which they do not.

It was against this backdrop that in 1992 the PFI was launched and almost immediately hit the rocks. The trouble came from two sides. First, the way in which civil servants had traditionally procured construction works and services left them without the experience, flexibility or negotiation skills to 'do deals' – a factor that was to prove such an important ingredient for advancement of the PFI. In addition, there was still a large divide and inherent suspicion between the public and private sectors, and very little guidance from government as to how this divide could be crossed. There is also little doubt that there was a faction within the public sector that would have liked to see this public/private separation maintained. Hence in 1993 the Private Finance Panel was created to encourage the use of the PFI, and further attempts were made in 1994 to ensure engagement of the public sector with the PFI when the then Chancellor, Kenneth Clarke, made it plain that Treasury approval for capital projects would not be given unless they had been tested against the private finance model.

The second major problem in trying to get the PFI off the ground related to the way in which, in the early days of the initiative, a whole range of projects were earmarked by overzealous civil servants as potential PFI projects when they were quite obviously not. The outcome of this was that consortia could spend many months or even years locked into discussions over schemes with little chance of success, because the package under negotiation failed to produce sufficient guaranteed income to pay off their debt due to onerous contract conditions and inequitable risk transfer stipulations by the public sector. This practice earned the PFI the reputation of incurring huge procurement costs for consortia and contractors before it became apparent that the business case for the project would not hold water. The procurement costs were non-recoverable by the parties concerned, and before long the PFI earned the reputation of being procurement of the last resort – at least by the private sector. In the mid-1990s the trade press ran several vigorous campaigns to revise and streamline the PFI, among reports that contractors had incurred costs of millions of pounds during abortive PFI negotiations. By October 1996, contractors Bovis, Taylor Woodrow, Tarmac and Trafalgar House abandoned PFI bids, citing bureaucratic delays. The 1997 Labour government was elected to power on a pledge to put partnership at the heart of modernising public services, and within a week of winning the election in May the Labour government appointed Malcolm Bates to conduct a wide-ranging review of the PFI. The first Bates Report made 29 recommendations, of which the dissolving of the Private Finance Panel in favour of a Private Finance Taskforce was key. The Taskforce had two wings; an advisory wing and a projects wing. The advisory wing developed a standard learning package for the public sector and helped greatly to disseminate information about the PFI procurement process, which until then had seemed something of a black art. Following the publication of the second Bates Report in 1999 and its recommendation that deal-making skills could be strengthened and that all public sector staff engaged in PFI projects should undergo annual training, PricewaterhouseCoopers were tasked with producing a PFI competence framework. The object of the report and the framework was to identify the competences required by the public sector in order to deliver successful PFI projects.

In June 2000 Partnerships UK (PUK) was established, following the publication of the second Bates Report (1997) in 1999. Partnerships UK replaced the projects wing of the Treasury Taskforce as a joint venture between the public and private sectors, with the private sector holding the majority 51% interest – it is itself a PPP! The mission of PUK is to supply expertise to the public sector in order to provide better value for money for PPPs. Included in its remit is the sourcing and provision of finance or other forms of capital where these are not readily available from established financial markets, and it makes a charge for its services. Most significantly, PUK can be seen to mark a move towards greater centralisation in the management of PPP projects and the development of standard documents, including contracts, in direct contrast to the mid-1990s, when each government department was encouraged to develop its own specialist expertise.

The government was also anxious to spread the use of private investment into local authorities, and in April 1996 the Local Authorities Associations established the Public Private Partnership Programme (PPPP, or 4Ps) in England and Wales. The 4Ps is a consultancy set up to help local authorities develop and deliver PFI schemes and other forms of public private partnership. The local authority services covered by the 4Ps include housing, transport, waste, sport and leisure, education etc. A more detailed explanation of the 4Ps process is to be found later in this chapter, including the criteria for the allocation of PFI credits.

The nature of the private finance initiative

The primary focus for the PFI to date has been on services sold to the public sector, and there are three types of PFI transactions currently in operation:

1. Classic PFI
2. Financially freestanding projects
3. Joint ventures.

Classic PFI

Services are sold to the public sector by the private sector, which is responsible for the upfront investment in capital assets. The public sector client pays only on delivery of the services to the specified quality standards. Examples of this type of project include Bridgend and Fazakerley Prisons and Debden Park High School, where custodial and educational services are provided by private consortia who build, finance, operate and maintain and, in the case of Fazakerley Prison, staff the facility. Figure 4.1 illustrates a typical contract structure for a classic PFI project. The scheme always remains in public sector ownership.

The classic PFI players

- *The special purpose vehicle/company*. A special purpose vehicle is a consortium of interested parties brought together in order to bid for a PFI project. If successful, the consortium

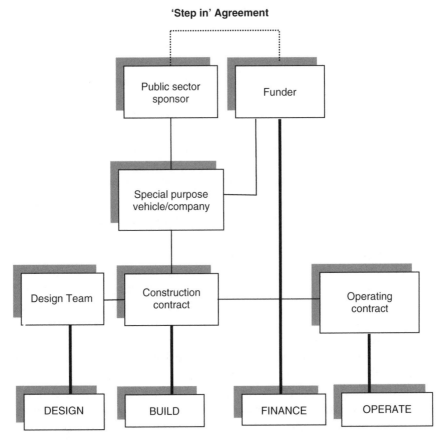

Figure 4.1 Classic private finance initiative – contractual relationships.

will be registered as a special purpose company, usually at the time of financial close. The special purpose vehicle/company, or 'shell company', is a unique organisation constituted purely for a single PFI project. The company has a contractual link with the public sector sponsor and the provider of finance, as well as the design, build and operating sides of the project. The funder also usually has its own agreement with the sponsor, which generally contains a step-in clause. This agreement is a safety net in the event that the special purpose company (SPC) ceases trading or persistently fails to deliver services to the required contract standards. In the event of SPC failure the whole project could be in jeopardy, as the supply of finance to the organisation running the operating contract would stop. The step-in clause permits finance to be channelled through the sponsors to ensure continuity in service delivery, which if interrupted, especially in the case of a hospital or school, would have wide-ranging consequences. In order to operate successfully, the special purpose company will have to call on the expertise of a wide range of consultants in, among other things, finance, design, construction, facilities management, service delivery and employment law.

- *The public sector sponsor.* This is the public sector client. Experience has shown that one of the key roles within the PFI procurement process is that of the client team's project manager. Optimum progress is made when the project is managed by a person who has the time and the authority to take decisions and negotiate with bidders, instead of having to keep referring back. The public sector client also usually relies heavily on input from consultants in the fields of procurement (including EU procurement law, see Chapter 8), project planning, production of an output specification, evaluation of bids, drawing up of contract documents and business cases, etc.
- *The funder.* In the early days of the PFI, raising finance was thought to be a major obstacle in the success of a project. Sourcing finance from the private sector is more expensive, and low returns on investment, unamortised debt and uncertainty of the implication of contracts that included the provision of words like *'force majeure'* made conservative financial institutions very reluctant to become involved. However, involved they must be if a PFI deal is to be signed, and at the earliest stage possible. In the final stages of the procurement, the most common problem facing purchasers and providers alike is the due diligence procedures carried out by the funders. These checks often pose serious questions about risk and other aspects of the contract that purchasers and providers think they have already resolved. Banks rarely commit the time of their own staff until they have decided that they have enough of an interest in the scheme to make it worthwhile for their own legal experts to go through the document in detail. This final step can be a trial of nerves, as the process can take several weeks or even months and can (and frequently does) involve previously agreed points being renegotiated to the satisfaction of the funders. As the PFI process has matured the attitude of the financial insti-

tutions has become less suspicious and they are generally more willing to fund projects. One of the results of this change in position is the reduction in the cost of funding and the increase in sources. Funding options now include: on balance sheet (i.e. from existing corporate funds); equity and subordinated debt from consortium members; leasing or similar tax-driven instruments; the bond market (either wrapped or unwrapped); senior debt, normally provided by major banks on the same basis as normal project financing; and mezzanine or junior debt, normally supplied by specialist departments or subsidiaries of major banks and by independent specialists. It is now thought by many financial institutions that the differential between public and private sector finance is almost insignificant – a major shift in attitude that has led to early PFI projects being refinanced. Table 4.1 illustrates the trends in PFI financing during the past 10 years or so. In particular, the reduction in return on equity indicates a process that is maturing from the initial high-risk proposition to mainstream financing.

- *The design and construction team.* The design and construct part of the process is usually the most straightforward and easily understood part of the procedure, with the majority of design teams and contractors leaving the project once the construction phase is completed and ready to start operating. One of the major criticisms of PFI projects has been their lack of architectural merit and design innovation. Some of the causes given are that the design period is too short, and that there is too much commercial pressure and insufficient contact with the user/client. One proposal to try to increase the quality in PFI design was tested on a £55 million law court project in Greater Manchester, where the design for the project was tendered for separately from the main deal. A shortlist of architect-led design teams is

Table 4.1 Private finance initiative trends in financing

	Early PFI	*Current market*
Term	18 years	25–30 years
Debt/equity ratio	80/20	92/8
Annual debt cover	1.25	1.15
Return on equity	20–30%	12–15%

submitting proposals for the project, which is expected to start in 2003 and be completed in 2005.

- *The operator.* The Operator is the 'O' in DBFO, and in the case of say, a school, this organisation will have complete devolved responsibility for the day-to-day operation of the facility, which may include such diverse services as the provision and maintenance of information technology, lunches, and playing field maintenance. In the case of ITCs there is often a contractual obligation to ensure that the hardware and software are kept up to date with the latest versions of programmes and technologies. The operator is clearly the major player is ensuring project success.

Financially freestanding projects

The second PFI model is one where the private sector supplier designs, builds, finances and then operates an asset, recovering costs entirely through direct charges on the private users of the asset (e.g. by tolling) rather than payments from the public sector. Public sector involvement is limited to enabling the project to go ahead through assistance with planning, licensing and other statutory procedures. There is no government contribution or acceptance of risk beyond this point, and any government customer for the specific service is charged at the full commercial rate. Examples of this kind of project include the second Severn Bridge, the Dartford River crossing and the Royal Armouries.

Joint ventures

Finally, joint ventures are where the costs of the project are not met entirely through charges on the end user but are subsidised from public funds. In many cases, the public sector subsidy secures wider social benefits not reflected in project cashflows (e.g. reduced congestion, economic regeneration). However, there could also be service benefits (e.g. from a shared facility or direct financial reward). The subsidy can take a number of forms, but the government role is limited to a contribution towards asset development. Operational control rests with the private sector. Examples include joint venture business park developments, various city and town centre regeneration schemes, Manchester's Metrolink, and the Docklands Light Railway Extension. This model is also used for PPPs.

Why the private finance initiative?

As mentioned in Chapter 3, certain public sector departments take the view that the PFI approach is now the preferred method of procurement. In the National Health Service, for example, it has become the dominant method; whether this move away from the diversity of procurement paths is wise is open to debate, but what has caused this shift? During 2000–2001, the Institute for Public Policy Research established a Commission on Public Private Partnerships to undertake the largest ever survey on the future of the provision of public services. The report was published in June 2001, and identifies a spectrum of approaches that exists for providing public services:

- Public sector default – here the public sector provides all services (e.g. clinical services).
- Private sector rescue – here the public sector provides all services, except if public providers are seen to be underperforming, in which case the private sector acts as a provider of last resort.
- Level playing field – here there is equal treatment between different organisations seeking to deliver public services; the decision regarding who is to provide the service is made solely on a judgment of which provider will give the best service (e.g. social services). This approach is consistent with plurality of service provision, and emphasises quality and value for money criteria over ideology. It also accords with the view of the private sector.
- Public sector rescue – here the private sector provides all services, except if private providers are seen as failing, when the public sector would act as provider of last resort.
- Private sector default – here the private sector provides all services on contract to public purchasers/commissioners (e.g. prisons).

At this stage of its development, it remains unclear why one approach applies in one sector but not another.

When launched in 1992, the mission of the PFI was stated by Chancellor Norman Lamont to be 'allowing the private financing of capital projects'. In a speech to the annual PFI Conference in October 1996, Chancellor Kenneth Clarke said:

> The injection of private capital investment and expertise that the PFI brings is a key factor in reconciling the need for sound public finances with acceptable levels of taxation and huge investment in infrastructure.

In 2001, Andrew Smith, the Chief Secretary to the Treasury, wrote:

> It is central to the government's approach to use the PFI/PPPs over traditional forms of procurement only where they provide better value compared to traditional public provision. But better value for money means that we can deliver more essential services and to a higher standard than otherwise would be the case.

Between the delivery of these statements there lies almost a decade of confusion and exaggeration about the rationale for PFI, with claims and counterclaims about its worth, as well as very real concerns about the ethics of allowing private organisations to build, operate and maintain the means of delivering healthcare and education for profit. What is also evident is the lack of an authoritative set of guidelines that identify the nature of the relationships between public and private sectors in PFI.

Since the introduction of the PFI in 1992, there has been a mismatch between the extravagant claims of its promoters and the more modest results achieved in practice. For example, in November 1995 the Private Finance Panel issued a list of projects identified for PFI consideration, including a £350 million project for the provision of a second Forth Road Bridge! No wonder therefore that claims were made in the media that the PFI was nothing more than a 'smokescreen behind which the Chancellor could take an axe to the investment budget'.

Whatever the confusion, the motive behind PFI remains the same; to realign the public sector from being an owner of assets to being a procurer of services. The government has a duty to provide and procure high quality public services in sectors such as healthcare, education, custodial services, etc. However, the government does not have any need to own, maintain or operate the built assets that are essential to provide these services. The PFI allows a licence to be granted to a suitably competent, privately owned consortium to build, finance, operate and maintain hospitals, schools and prisons, in return for which the consortium will receive a guaranteed payment for the duration of the service contract – typically 25 years minimum. The services must be delivered to predetermined standards and benchmarks – for example, National Health Service or Prison Service standard procedures. At the end of the contract term the facility (in the case of a prison, for example) is handed back, at no cost, to the Home Office in a well-maintained condition. The prison service has had the benefits of a modern, well-equipped, well-maintained and staffed prison for 25 years, and in return the private consortium has received a guaranteed income over the same period. Initial enthusiasm for the PFI was perhaps bordering on a 'something for nothing' mentality – that is to say, instead of the government paying £65 million upfront for a new prison, it was much better to get the private sector to provide, maintain and operate it. However, there truly is no such thing as a free lunch, and it should not be forgotten that the PFI leaves a legacy in the payments that must be made under long-term contractual arrangements by successive governments. Table 4.2 illustrates the estimated payments that must be made by the Treasury under existing PFI contracts until the year 2019, although contracts have been signed committing successive governments to make payments until 2031.

In Chapter 1 it was noted that the Royal Institution of Chartered Surveyors, in its 1998 report *Challenge for Change*,

Table 4.2 Payments committed to private finance initiative projects

	£ million		£ million
2001–2002	2900	2010–2011	3500
2002–2003	3600	2011–2012	3600
2003–2004	3600	2012–2013	3600
2004–2005	3600	2013–2014	3400
2005–2006	3500	2014–2015	3400
2006–2007	3700	2015–2016	3100
2007–2008	3700	2016–2017	3200
2008–2009	3600	2017–2018	3200
2009–2010	3500	2018–2019	2700

highlighted the PFI/PPP as a major factor in determining the future role of the quantity surveyor. Public private partnerships would appear to be rampant – PFIs dominate the debate on the provision and delivery of public policy, from healthcare to education to housing. The reality is somewhat different; the PFI could be abolished tomorrow and all public investment financed conventionally without breaking the Chancellor's current fiscal rules. Indeed, over the period covered by the Labour government's first comprehensive spending review (1999–2002), the PFI is planned to account for only 12–14% of total publicly sponsored gross capital investment. Treasury officials tend to emphasise the modest scale of the PFI/PPPs, although not so ministers. Consequently the technical press and the wider media announce almost daily new deals or partnerships to provide new public sector services. Table 4.3 illustrates the estimated amounts of PFI and capital spending over a 4-year period to 2003 and shows that, in comparison to overall public sector expenditure, the PFI input is comparatively small, falling to a forecast 10% by 2002–2003.

By December 2001, agreements for over 400 private finance initiative projects had been signed by central and local government for the procurement of services across a wide range of sectors, including roads, railways, hospitals, prisons, office accommodation and IT systems. Of these, 180 projects have been completed, with a capital value of £5.1 billion. Of the 400 agreed

Table 4.3 Projection of PFI capital spending

	1999–2000	*2000–2001**	*2001–2002**	*2002–2003**
Total public sector expenditure including PFI (£bn)	19.8	26.6	31.1	34.9
Estimated PFI capital expenditure (£bn)	1.6	4.3	4.4	3.6
Percentage of PFI	8.0	16.0	14.0	10.0

*Forecast figures.

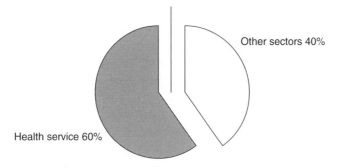

Other sectors 40%

Health service 60%

Figure 4.2 Agreed PFI deals.

deals, 179 (148 in England, 6 in Wales and 25 in Scotland) are in the health service (Figure 4.2).

In monetary terms, the health sector is a major PFI player.

The current state of the private finance initiative

So what is the current state of health of the PFI, and why is it used? When it was first launched in 1992, the principal rationale was to provide value for money and efficiency savings, as well as to transfer risk from the public to the private sector. This remains pretty much unchanged today, and will be discussed in more detail later. However, over the past ten years there have been claims by both opponents and proponents that the PFI also:

1. Makes more investment in key public services possible
2. Helps reconcile high capital spending with prudent fiscal policies
3. Is a smokescreen to enable the Treasury to take an axe to public investment
4. Saves money by delivering cheaper solutions
5. Fosters innovation in design and service delivery.

Looking first at claims 1–3, the accusation has been made that the PFI is in some way a Treasury conjuring trick to balance the public expenditure books. In the past decade a number of changes in fiscal policy and government accounting procedures have all had an effect on the PFI development, including the introduction of 3-year comprehensive spending reviews to provide departments with greater certainty about future spending allocations. In 1998 the government introduced new public sector financial control networks that not only renamed the PSBR as the public sector net cash requirement (PSNCR), but also greatly downgraded the significance of this once important fiscal measure to the point where it is no longer even referred to by the Chancellor in budget statements. In its place two new rules were introduced; the so-called 'golden rule' (over the economic cycle, the government will borrow only to invest and not to fund current spending) and the 'sustainable investment rule' (public debt as a proportion of national income will be held over the economic cycle at a stable and prudent level).

Without going into too much detail, an implication of this change was to provide a more conducive regime for public investment, which allowed for an increase in government borrowing for capital projects. Call it creative accountancy if you will, but it would now be possible for government to finance all planned PFI spending directly through traditional methods and still remain within the government's medium-term fiscal objectives. Clearly, this being the case, claims 1–3 no longer have any credence.

Turning now to claim 4, does the PFI in fact provide cheaper projects? The National Audit Office has produced several reports into the savings achieved by the PFI, and Table 4.4 summarises their findings.

One of the most high profile sectors that has adopted PFI is the National Health Service where, conversely, the savings

Table 4.4 PFI savings compared (source: National Audit Office)

Scheme	*Present value PFI cost (£m)*	*Saving compared to traditional procurement (public sector comparator) (£m)*
First Four Roads	698	99
A74/M74	193	17
Bridgend and Fazakerley Prisons	513	54
Dartford and Gravesend Hospitals	177	5
NRS2	134	196
Prime	1986	560
Total	3701	931

have been minimal, typically less than 5%. It has been argued that the decision not to transfer responsibility for clinical services to the private sector means that the private sector has to recover its capital investment from only a relatively small part of the operation's cashflow; in prisons, on the other hand, full operational responsibility has been transferred and efficiency savings of up to 10% achieved. As illustrated in Table 4.4, the average estimated saving on a sample of PFI projects was 17% when compared with traditional procurement. However, if NRS2 and Prime, two ITC projects, are excluded, the savings are reduced to around 10%, which in monetary terms converts into future savings on public sector capital projects of around £400 million a year by 2005. To place this in the context of total public sector spending, consider that the planned spend in 2002 on the National Health Service is expected to be £46 billion, and on Social Security £102 billion. Clearly, therefore, although PFI does deliver public services at reduced cost, the savings are marginal and PFI contributes less than is claimed to saving public money. Consequently claim 4 – that the PFI delivers cheaper solutions – can be said to be not proven. In addition, the Arthur Andersen/Enterprise LSE report (2000) *Value for Money Drivers in the Private Finance Initiative* came to the conclusion that projected savings are sensitive to and in some cases totally dependent on risk transfer valuations, which is a particularly problematic area to evaluate.

Finally, this leaves the claim 5 – that the PFI encourages creativity and innovation in design and delivery of services, which will be achieved only if a private sector bidder can produce a solution that will reduce costs and/or result in an asset better able to deliver a scheme's service requirements. The emphasis on ongoing service delivery and whole life costs generates a number of crucial benefits as the contractor now has a much stronger incentive to ensure fitness for purpose of the built asset over the life cycle of the contract. No cones on PFI roads a few days after opening! However, a major constraint on innovation has been identified as the planning process; the same restrictions and legislation apply as with any other project, and in most cases it is claimed that the design period is too short. In the majority of cases, therefore, innovation would seem to come from the successful integration of design, construction and operation – the 'joined up' process so sought after by clients.

Hence it would seem that this factor does have a significant contribution to delivering added value.

It would appear therefore that the original drivers to provide added value (however measured) and innovation in service delivery are still the main focus of PPP/PFI, and that the transfer of risk is also a crucial factor in tipping the value for money balance, especially in light of the magnitude of the savings shown in Table 4.4. There can be no doubt that PPP and the PFI are here to stay, and their strategic importance in the provision of high quality public services is much more than creative accountancy. What is more debatable is the influence that PPP/PFI will have on the long-term direction of the UK construction industry and the future of the quantity surveyor. The Institute for Public Policy Research Commission report (2001) suggested that the building blocks that support the use of PPP are as follows:

- The problems that modern government faces cannot be solved by government acting in isolation. Public authorities increasingly recognise that by working with others they are better placed to deliver the policy outcomes they want, whether these are high quality healthcare or efficient street lighting.
- Government is not the primary source of social and economic problems, and improved outcomes cannot be achieved simply by restricting its activities. Out-and-out privatisation is therefore not the answer to alleged shortcomings in public services.
- Public services and policy makers need to be continually striving to find new ways of improving the quality of the services that they provide. Ever rising public expectations, advances in technology and innovations in potential delivery mechanisms make clinging to the *status quo* no longer viable.

The Commission also came to the conclusion that PPPs are not a panacea for resolving all the challenges that modern government faces. They may be an important part of a wider strategy for the revitalisation of public service, but they are not a way of resolving how public services should be funded. The National Audit Office report *Examining the Value for Money of Deals under the Private Finance Initiative* (National Audit Office,

1999) suggested that the four pillars of getting a good PFI deal are:

1. Setting clear objectives – procuring departments need to think through, in advance, exactly what they are looking for from the proposed deal, and how that outcome can be delivered
2. Application of the proper procurement processes – the process must comply with the relevant law and regulations and must be designed to maximise the prospect of achieving a deal that is good value for money
3. Getting the best deal available – the prime focus should be on the quality of the bids received
4. Ensuring that the deal makes sense – the terms that were envisaged at the outset of the project should be maintained throughout the procurement process, and the project should be reconsidered if these terms are not met.

The PFI procurement process – 'getting a good deal'

Table 4.5 illustrates the recommended Treasury Taskforce procurement path for PFI projects. The Pricewaterhouse-Cooper report *PFI Competence Framework* suggested that these stages fall into three broad phases (Pricewaterhouse Cooper, 1999):

1. Phase (i), Feasibility – Stages 1–4
2. Phase (ii), Procurement – Stages 5–13
3. Phase (iii), Contract management – Stage 14.

NB: Although the Treasury Taskforce no longer exists, the data produced by this organisation are still recognised as being valid and appropriate.

This section of the chapter will concentrate on the stages that have proven to be of crucial importance in determining the case for the use of the PFI, namely:

- Stage 3 – the assessment of value for money and risk transfer The Outline Business Case
- Stage 9 – Refining the proposal
- Stage 12 – Selection of the preferred bidder.

Table 4.5 PFI procurement guide (source: HM Treasury)

Stage 1	*Establish business case.* It is vitally important that the PFI project is used to address pressing business needs. Consider key risks
Stage 2	*Appraise the options.* Identify and assess realistic alternative ways of achieving the business needs
Stage 3	*Outline business case.* Establish the project is affordable and 'PFI-able'. A reference project or public sector comparator should be prepared to demonstrate value for money, including a quantification of key risks. Market soundings may be appropriate at this stage (see Chapter 7). The outline service specification should be prepared
Stage 4	*Developing the plan.* Form procurement team with appropriate professional and negotiating skills
Stage 5	*Deciding tactics.* Identify the nature and composition of the tender list and selection process
Stage 6	*Publish OJEC.* Publish the contract notice in OJEC (see Chapter 7 and Appendix)
Stage 7	*Prequalification of bidders.* Bidders need to demonstrate the ability to manage risk and deliver service
Stage 8	*Selection of bidders.* Shortlist bidders. Method statements and technical details may be legitimately being sought
Stage 9	*Redefine the proposal.* Revisit the original appraisal (Stage 3) and refine the output specification, business case and public sector comparator
Stage 10	*Invitation to negotiate.* This stage could include draft contracts and is quite lengthy (3–4 months). Opportunity for shortlisted bidders to absorb contract criteria and respond with a formal bid
Stage 11	*Receipt and evaluation of bids.* Assessment of different proposals for service delivery
Stage 12	*Selection of preferred bidder.* Selection of preferred bidder with bid being tested against key criteria
Stage 13	*Contract award and financial close.* Sign contract and place contract award notice in OJEC (see Chapter 7)
Stage 14	*Contract management.* Operational and management relationship between public and private sectors

Stage 3 – The Outline Business Case

The outline business case is at the heart of the feasibility phase. Assuming that the need for the project has been established and other methods of delivery considered, the preparation of the outline business case is the starting point for an audit trail that runs through the procurement process and should contain the following:

1. Identification of key risks
2. Output specification
3. A reference project or public sector comparator.

Identification of key risks

Risk transfer is one of the key tests for a good PFI deal, as value for money can be demonstrated to increase each time a risk is transferred. There are two aspects to risk transfer:

1. Transfer between the public and private sectors
2. Transfer between the members of the PFI consortium.

In most PFI projects the risks that are earmarked for transfer to the private sector are by now fairly standardised and well understood; however, major difficulties arise in deciding who within the consortium carries the various burdens of risk. This factor would seem to explain why complete teams, which include both contractors and facilities management operators, are increasingly successful at winning PFI bids. In the case of risk transfer between public and private sectors, the main drivers are transparency and the need to demonstrate value for money; in the case of risk transfer within the consortium, the commercial interests of the various players (i.e. financial institutions, contractors, operators, etc.) dominate the discussion.

The principle governing risk transfer is that the risk should be allocated to whoever from the public or private sector is able to manage it at least cost – that is, identified risks should be retained, transferred or shared. The valuation of risk transfer, however problematic, often tips the scales on PFI deals as the public sector comparator alone often shows that value for money has not been demonstrated. In 6 of the 17 projects analysed in the Arthur Andersen/Enterprise LSE the assessment of value for money was entirely dependent on risk transfer valuation, and in all 17 projects it accounted for 60% of all cost savings, making this element of the deal particularly important. There will always be a wide variety of risks associated with potential PFI projects, including the following.

Risk transfer between the public and private sectors

- *Design and construction risk.* The construction period of a PFI project is recognised as one of the most critical phases and for this reason often attracts the highest valuation, which in some cases can account for 50% of all risk valuation. If for any reason the project is not complete or is late and the service, whether it is healthcare or education, cannot be delivered, the income stream will not be generated. Generally, the design and building of the asset is a risk best borne by the private sector consortium and its financial backers. This is because the built asset is being designed to an output/service specification instead of to a rigid set of departmental guidelines, and there is a commercial incentive for efficiency right through from initial design to build and operation.

- *Commissioning and operating risks* (including maintenance, or whole-life costs). It has been estimated that over the 25-year life of a PFI contract, on average 35% of all costs will be capital cost while 65% will be running and maintenance costs. In fact in some projects the split could be a great as 20/80%. The golden rule for consortia is therefore 'concentrate on the large'. As discussed in Chapter 1, the UK construction industry has traditionally ignored the influence of whole life costs in building design; the PFI ensures that they are ignored at the consortium's peril. Once again, this risk is best managed by the private sector.
- *Whole-life costs.* Whole-life costs are thought by many to lie at the heart of the PFI, particularly as such a high percentage of costs associated with a PFI project are to be found in running and maintenance costs, and every public sector comparator is built upon the basis of whole life costs. This makes sound commercial sense in the design, build, finance and operate context, where risk is being transferred over a long time period and has to be priced. Whole life cost evaluation has been an established procedure in some sectors of the construction, oil, gas and engineering industries since the mid-1960s. Over the past 40 years the term to describe the technique has changed from costs-in-use to lifecycle costing to terotechnology to total ownership costs. Whatever the term used, the basic technique remains the same; it is a process that allows the comparison of various alternative design solutions using discounting techniques by taking into account not only the initial capital cost but also considerations such as running costs, operating costs, maintenance costs (both annual and planned) and capital allowances and

tax considerations over the anticipated lifespan both of individual components and of the capital project as a whole. Despite the obvious advantages of considering whole life costs, in much of the private sector this has yet to become a widely used decision support tool. In a combined DETR and BRE report, the reasons identified for the lack of enthusiasm include: significant technical barriers, including lack of data on both costs and performance; and motivational barriers, including lack of client interest and trust. The critical difference between the PFI and typical private sector procurement is that the knowledge, expertise and control over life cycle risks are in the hands of the service provider, who has a major incentive to optimise. Unless the lifecycle risks are managed, the price for the job will be wrong. Often neither the client nor the bidder has been able to pull together all of the necessary information, which perhaps explains why large organisations such as the French giant Bouygues, with its own in-house facilities management expertise, are members of consortia winning PFI deals.

- *Demand (volume) risk.* The prospect of receiving a stable long-term income flow is a major attraction for many private sector consortia that bid for PFI projects. Demand or volume risk is uncertainty regarding the level of demand for the service provided by the consortium-operated asset. Generally the private sector is unwilling to accept volume risk, as it is usually the public sector that has control over volume – for example, the numbers of convicted prisoners requiring a place in custody, or the numbers of patients requiring a hospital bed, both of which are units of payment.

- *Residual value risk.* Residual value risk is uncertainty over what the net value of the asset will be at any time during the contract period. It is highlighted as an issue when a PFI scheme anticipates the transfer or sale of the asset and requires case-by-case consideration. For instance, the procuring entity may no longer require the asset at the end of the service contract. There are two main determinants of residual value; first, the condition of the asset at the end of the contract; and secondly, the demand (if any) for the asset. Some projects, such as prisons, obviously have limited possible alternative uses; however, the private sector takes responsibility for maintaining the asset in good condition.

- *Technology and obsolescence risks.* Technology risk is associated with the obsolescence of both the services and the function of the assets themselves. It is generally not thought to be significant outside IT projects; however, it would be a brave person who tried to predict methods of healthcare delivery 20 years from now. Technology refreshment, for example replacing computer networks within a school every 5 or 10 years, is a risk usually transferred and managed by the private sector, but wider-ranging obsolescence risks impacting on the mode of service delivery would be a matter for negotiation.
- *Legislation risks – both UK and EU.* Legislation or regulatory risk is one special aspect of demand risk, and is thought to be outside the influence of the private sector. An example would be a reduction in the resources available to the NHS due to legislation passed by the UK government, the Scottish Executive or the European Parliament. The test should be one of materiality, with the risk of general changes in the law being borne by the private sector, although the way should be open to negotiate possible price adjustments in the event of changes with a major impact.
- *Project finance risk.* Project finance risk is the risk associated with the ability to raise finance on the terms suggested by the consortium in their bid. It is a risk retained by the private sector. One of the major criticisms of the PFI was that using private sector finance would always be more expensive than public sector funding at 6% plus inflation. The maturing of the PFI now means that the gap has narrowed to a point where it is almost insignificant.
- *Termination risk.* Most PFI contracts include a statement of the circumstances that would result in the contract being terminated. Generally, termination occurs if the consortia fail to provide the level of performance statement in the service specification. Under these circumstances, the public sector would be exposed to taking over the running of the service and paying off any debt associated with the project.
- *Refinancing risk.* The National Audit Office now recommends that risks associated with the refinancing of PFI deals be considered at an early stage. Refinancing (explained below) can result in the public sector being exposed to an increase of termination risk – that is, additional debt that has to be repaid in the case of the contract with the private company being terminated.

Refinancing and clawback

Refinancing is the process by which the terms of the funding that was put in place at the outset of a PFI contract are later changed during the currency of the contract, usually with the aim of creating refinancing benefits for the consortium company. There are benefits to shareholders of increasing and/or bringing forward their returns from the project as a result of changes to the financing structure of the consortium company. Refinancing is particularly attractive for projects that went to financial close in the early days of PFI, when financial packages usually carried a high price tag.

For example, in November 1999 Fazakerley Prison Services Ltd refinanced the project that it had been awarded in 1995 to build, maintain and operate Fazakerley prison for 25 years. The project, the first PFI prison, was negotiated at a time when PFI was in its infancy, and consequently the financial package that made it possible reflected the banker's perceived high risk of this new procurement route. Refinancing was able to proceed in 1999 because of the successful track record of the company in operating the prison, and also because the increasing levels of confidence in the financial markets towards PFI projects generally meant that the financial markets had scaled down their risk assessment rating. The refinancing improved the expected returns to the shareholders, both through the early payment of the original investment and by generating a more favourable flow of dividends. These changes increased returns by £10.7 million and, when added to the £3.4 million bonus already paid for early delivery of the prison at lower than expected cost, boosted the expected returns by £13.1 million (61%), from £17.5 million to £30.6 million, taking into account £1 million repaid to the prison service for additional risk associated with the refinancing deal. The original PFI contract with the Fazakerley consortia did not give the prison service any automatic rights to claw back or share in the benefits of higher than forecast profits, although the prison services permission was sought before the deal was refinanced. This is clearly a contentious issue, as the private sector could be accused of making excess profits via the back door from the delivery of public services, and for some it strengthens the case against the PFI. As a direct result of the Fazakerley deal, the National Audit Office published a report in June 2000 that set out general principles that government departments should apply to refinancing, *viz*:

- Appropriate benefits should go to those bearing risks
- Benefits from reducing costs in a developing market should be shared if they have not previously been reflected in the contract price
- Departments that sponsor refinanced projects should seek compensation for any exposure to increased risk
- If the private sector seeks refinancing then it is reasonable for the public sector to seek a share of the benefits.

The report also pointed out that substantial refinancing gains may threaten public perception of value for money.

The degree of transfer of other risks is a matter for negotiation at project level. In addition to the NAO's report, in July 2001 the Office of Government Commerce issued guidance for government departments when refinancing PFI projects in which it urged caution on departments and suggested that, with regard to any reallocation of risk, any benefits from refinancing should be shared 50/50 with the private sector partner.

Assessing risk

The value of risk transfer, both quantitative and qualitative, is a vital part of the assessment of value for money (VFM) in the public sector comparator (PSC) of a PFI project. The process involves the following stages:

1. Identification of risks
2. Assessment of the impact of risks
3. Assessment of the likelihood of risks arising
4. Adjustment of the financial model
5. Allocation of risks.

For a large PFI project the identification of risk is likely to be a long and complex procedure, and the interaction between various identified risks must also be considered. The allocation of risk with the preferred bidder will be the subject of negotiation. However, for the purpose of establishing the feasibility of the project and the PSC the following assumptions have been made, in this case based on previous experience.

Table 4.6 Risk register

Commercial risk	Purchaser	Operator
Demand risk	(£10m)	
Third-party revenues		£20m
Design risk		(£25m)
Maintenance risk		(£20m)
Obsolescence		(£25m)
Residual value	£10m	

- *Stage 1 – allocate risk.* Table 4.6, the risk register, lists some of the risks that have been identified for a prison project, together with an initial valuation – the figures in brackets represent possible losses.
- *Stage 2 – estimate commercial significance of risk.* The figures in the risk register are drawn either from empirical evidence or, where that is not available, from commonsense estimates based on specialist knowledge. As the assessment of risk transfer is such an important part of proving value for money, the figures should be as accurate and as up to date as possible.
- *Stage 3 – assessing the likelihood of risks arising.* Having established the value of the risk in net present value terms, the next step is to calculate the probability that everything will go according to plan and that, for example, the demand risk is correctly valued at £10 million. The most accurate and widely accepted method of doing this is by using a simulation modelling technique, of which there are many, but perhaps Monte Carlo simulation is the most respected. In order to apply this technique it will be necessary to prepare an estimate of the probability distribution within upper and low limits. After applying Monte Carlo simulation to the demand risk estimate, it is shown that there is only a 70% probability that demand for the new facility will decrease to the point where the purchaser would lose money, and therefore the estimate can be downgraded to £7 million. The quality and accuracy of the information used to construct the simulation model is of crucial importance. This exercise is then repeated for every risk identified by the purchaser.
- *Stage 5 – allocation of risk.* Who is now going to accept the risk? The approach taken by some purchasers to allocation

of risk is to issue a risk matrix to the shortlisted consortia at Stages 7/8 in the procurement process, marking those items that are clearly not negotiable and are to be transferred but requesting shortlisted consortia to indicate which of the remaining risks they are prepared to accept.

Risk transfer within the consortium

As discussed previously, a consortium is a unique collection of organisations with very differing expertise, each attempting to achieve maximum return combined with minimum exposure to risk. One scenario for trying to achieve this is with the use of a financial model where factors such as desired return, whole-life costs and capital costs (generally based on costs/m^2 of gross floor area) are input, thereby determining what's left to compensate for the risk. This figure is then divided, after considerable debate among the various consortium members.

Interestingly, in a report published in November 2001 by the National Audit Office and entitled *Managing the Relationship to Secure a Successful Partnership in PFI Projects* (National Audit Office, 2001), only two-thirds of contractors shared the authorities' view that risks had been allocated appropriately; 79% of authorities thought the risk allocation was totally satisfactory, but only 53% of contractors held this view (Figure 4.3).

Output specification

The preparation of the output specification is one of the most important phases in the PFI bidding process. It is vital that the output specification states the core requirements in terms of output – that is, what is wanted, not how it is to be provided – thereby allowing the consortia the maximum opportunity for innovation in service delivery. The key elements of an output specification are seen as:

1. *Objectives*. There must be a clear statement of the strategic objectives of the project, expressed if possible in terms of delivery of service rather than the built assets.
2. *Purpose*. There must be a summary of the desired outputs of the project; for example, a new PFI hospital project may describe its purpose as follows:

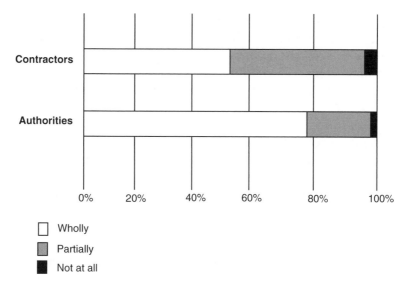

Figure 4.3 Perceptions of risk allocation (source: National Audit Office, 2001).

This project is to provide healthcare services for
- Thirty long-stay beds for elderly people
- Nineteen acute GP beds
- Out-patient services, including specialist input
- Day surgery and minor injury clinic.

The service will include the provision of catering, cleaning, hairdressing, transport and porterage (including ambulances), and security, together with arrangements for the maintenance of patient records.

3. *Scope.* The purpose of this section is to identify whether the proposed project has the potential to produce alternative income streams – for example, could a school building be used during weekends, holidays etc., or could a hospital be extended to include the provision of private accommodation? Who is to benefit from any extra income stream that this additional use may attract? Market sounding may be appropriate to ascertain the potential for the type of action. Clawback clauses are inserted into PFI contracts to share extra profits between the operator and the public sector clients. When no such clauses exist, the PFI consortia share out the profit among themselves, which some

argue is justified given the risks involved. For example, as discussed previously when the special purpose company restructured their debt for Fazakerley Prison after the financial close, it resulted in a near doubling of profit from 17.5 million to 30.6 million. Government guidelines are expected to be published covering this contentious issue.

4. *Performance.* The required performance levels of the project should be set out by way of operating outputs, without reference as to how this performance will be achieved. Importantly, the performance of the operator should be measurable and able to be evaluated, as payment will be directly related to this. Government agencies like the National Health Service and the Prison Service have standard criteria in areas such as waiting times for hospital admission, facilities and timing of recreation periods within a prison, etc., and these should be maintained by the private sector operator as a minimum level of service provision. One of the clouds hanging over the PFI is the issue of what happens if an operator fails to deliver a service to the required levels, as has happened in at least one case, when the public sector had to step in to take over the service delivery.

5. *Compliance and compatibility.* In the situation where a private operator is to provide a service within the framework of a larger service delivery, it is fundamental that the service being provided under the PFI project is compatible with the overall systems. One obvious example is the formats of IT systems.

6. *Constraints.* If the project is the subject of constraints, for example in terms of planning permission, then these should be made explicit.

7. *Risks.* The type and nature of risks should be set out as discussed above.

8. *Alternative solutions.* Tenderers must be allowed the opportunity to offer alternative solutions, without being prescriptive as to the nature of what these alternatives may be – for example:

Suppliers will be responsible for the maintenance or replacement of electrical and mechanical equipment throughout the life of the contract to a standard that permits service standards to be met.

Most PFI deals currently being negotiated will run for 25–30 years at least, and one of the major problems to be addressed when considering outputs is the way in which demands on the service may alter during that period. For example, it is particularly difficult to plan for healthcare delivery against the backdrop of an increasingly aging population (the number of people in the UK aged over 85 increased three-fold between 1971 and 2000, from 400 000 to 1.2 million) and the greater expectations of a consumer society for high quality care. There is the emergence of new technology, surgical practices, diseases and cures. In order to allow for change, purchasers should specify that the design of assets allows for some operating flexibility, and contracts should permit changes to the unitary charge arising from service changes.

Reference project or public sector comparator

Public sector comparators (PSC) incorporate a public sector reference project that provides a snapshot of value for money at a particular point in time. Put simply, a PSC poses the question, which gives better value for money; the conventional public procurement route or the PFI? The lack of transparency in the construction of a PSC has been criticised, and indeed there has been the accusation that figures used in this pivotal VFM exercise have not been scrutinised sufficiently rigorously. A PSC (or reference project) may be defined as a hypothetical risk-adjusted costing by the public sector, as a supplier, to an output specification produced as part of a PFI procurement exercise, and:

- It is expressed in net present value terms
- It is based on the recent actual public sector method of providing that defined output (including any reasonably foreseeable efficiencies the public sector could make)
- It takes full account of the risks that would be encountered by that style of procurement.

In order to be a valid benchmark against which private sector bids can be fully and fairly compared, the PSC must reflect not only certain procurement costs but also the additional costs of risks that may arise, and which under PFI would fall to the consortium. During the PSC process, risks should be identified and the cost impact evaluated (Figure 4.4).

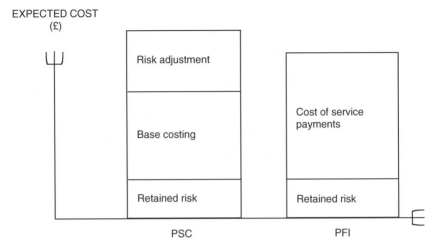

Figure 4.4 Establishing value for money with the public sector comparator (source: HM Treasury, 1998).

The format of a PSC is not fixed. However, the Treasury Taskforce has set out a recommended list of items that should be included (HM Treasury, 1999). These are:

- An overview and short description of the project
- An estimate of basic procurement costs and operating costs
- The opportunities for third-party revenues
- An estimate of the value of the asset on transfer or disposal
- A risk matrix showing their costs and consequences
- A discounted cash flow of future costs
- A sensitivity analysis showing the consequences of variance in key assumptions.

Stage 9 – revisit and refine the original appraisal

The outline business case is now complete and, assuming that VFM has been established, the procurement procedure can proceed. The figures prepared in Stage 3 will be revisited at Stage 9 before selected bidders are asked to submit their proposals and enter into negotiations. The original output specification should be refined to a set of core deliverable outputs, and a set of proposed key contractual terms and allocation of project risks prepared. Having selected a shortlist of prequalified bidders earlier in the process, the documentation is now ready to distribute during Stage 10.

Stage 12 – selection of preferred bidder and final evaluation

It is not uncommon for Stages 10 and 11 to take several months. On selection of the preferred bidder the PFI project should again be tested against the critical success factors of value for money, including comparing the preferred bid with the public sector comparator.

Concern has been expressed regarding the value of the PSC, as it is felt by some departments that, as PFI is the only procurement path likely to get approval, the degree of scrutiny of the figures used to compile it could be more rigorous. In an attempt to introduce more objectivity into the process, from January 2001 the Gateway Process will be mandatory for all new high-risk procurement projects and all IT procurement projects in central government, executive agencies and other non-departmental public bodies. The Office for Government Commerce introduced the procedure in order to apply an independent assessment and further scrutiny of the project being considered. The process is carried out by a team of between two and five people, depending on the nature of the project, and there can be up to five reviews during the life of a project, usually taking place over a 3–5-day period. The recommended timing of the reviews is shown in Figure 4.5.

A key feature of the PFI process is that it takes a long time to arrive at an agreed deal. Figures produced by the Department of Health indicate that from the publication of the OJEC notice to financial close was 46 months, while the Private Finance Unit of the Scottish Executive estimate that for PFI schools projects the period is between 98 and 116 weeks. PFI projects are complex; they are not a quick fix to an investment need, but the process should become quicker as parties become more familiar with this form of procurement.

PPPP and Local Authorities

Before leaving the subject of public private partnerships, the increasing role of this method of procurement at local government

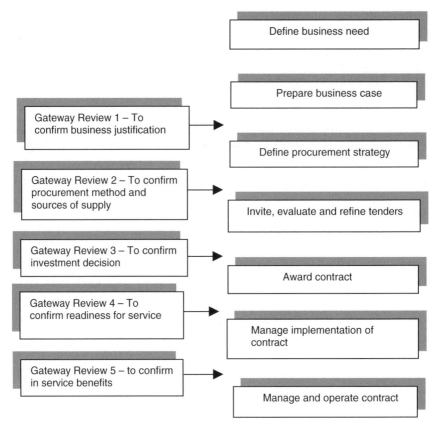

Figure 4.5 Gateway reviews (source: Office of Government Commerce).

level will be briefly discussed. There have recently been a series of high profile public private partnership projects at local level, particularly in the provision of schools. Stoke-on-Trent was the first local education authority to sign a deal, worth £153 million, to upgrade and maintain the entire stock of schools (amounting to 122) in the city, in a project awarded DfEE pathfinder status. Among the key issues for the local authority was the need to produce a scheme that improved the environment in schools as part of the drive to raise standards of achievement. Also, there was a desire to replace the crisis management response to fabric and plant failure with a maintenance regime based on planned and whole life considerations. The PPPP route for schools is illustrated, but it should be noted that each area of local authority service provision has to apply through the appropriate department or Ministry (for example, hospitals to the Department of Health, prisons to the Home Office, etc.), although the method of approval remains substantially the same.

Local education authorities wishing to develop PPPP schools schemes have to bid for PFI credits approval or borrowing permission, which are allocated on the basis of agreed priorities. PFI credits cover the capital elements of PFI schemes, and are similar to mainstream credit approvals in that they bring with them a stream of revenue funding through which local authorities meet debt charges relating to the repayment of capital investment. PFI credits must be secured before the local authority can go to the market. The amount of DfEE credits earmarked as available in 2002 is £350 million out of a total sum of £3.2 billion for all PPPP projects during the period March 1999–2002. Table 4.7 illustrates the income flow of PFI credits that could be available for a typical school project; the credit contributions have been discounted to net present values using a rate of 7.9%.

The agreed criteria for allocation of PFI credits to DfEE schools PPPP projects are for schemes that demonstrate:

- The extent to which the project will contribute to raising educational standards and tackle the most urgent premises-related needs
- The extent to which the project contributes to wider objectives, for example addressing social exclusion, provision of services for the wider community
- The extent to which the scheme demonstrates value for money and further develops PPPP in the school sector by innovation in service provision.

In all other respects the method of procuring the best value deal is exactly the same as previously described in PFI deals, and follows the procurement path detailed in Table 4.5.

Table 4.7 is based on the provision of a replacement building that has an 18-month construction period; support does not commence until the school is complete and operational. Note that further adjustments to the level of support will be carried out to take account of local authority accounting procedures, such as revenue abatement, which may result in a slightly lower figure of contribution.

Table 4.7 Calculation of the private finance initiative credit support

Year	Payments to the project company by LA (£)	PFI credit cover (NPV) contribution (£)
1	0	0
2	0	0
3	666 667	572 619
4	1 007 500	802 012
5	1 015 056	748 866
6	1 022 669	699 243
7	1 030 339	652 907
8	1 038 067	609 642
9	1 045 852	569 244
10	1 053 696	531 523
11	1 061 599	496 302
12	1 069 561	463 414
13	1 077 583	432 706
14	1 085 664	404 033
15	1 093 807	377 260
16	1 102 010	352 261
17	1 110 276	328 918
18	1 118 603	307 122
19	1 126 992	286 771
20	1 135 445	267 768
21	1 143 960	250 024
22	1 152 540	233 456
Total	21 157 886	9 386 091

The role of the quantity surveyor in PPP/PFI

There is the belief in certain quarters that PPP is a missed opportunity for the chartered surveyor. Other professionals seem to have grasped the nettle, while for many surveying practices involvement in PPP stops at the preparation of bills of quantities for PFI consortium contractors. As previously noted, the Royal Institution of Chartered Surveyors is in no doubt as to the importance of PPP/PFI in the future of the profession, and many quantity surveying practices are involved in PPP/PFI deals in a variety of roles for both the public and private sectors. In the private sector, working for the operator, the quantity surveyor role involves (note that all references are to Table 4.5):

1. *For the private sector – special purpose companies*:
   - Advice on procurement. For many private sector consortia, the approach to submitting a bid for a PFI project is unknown territory. Added to this is the fact that by their nature PFI projects tend to be highly complex, requiring decisions to be taken during the development of the bid at Stage 8 that involve both capital and long-term costs. Increasingly, the impact of EU procurement directives must be considered. Some contracting authorities use the OJEC to 'test the water' for a proposed PPP project at Stage 6; the quantity surveyor can supply preliminary cost information at this time. In addition, the quantity surveyor with experience in PPP can provide expert input into the pre-qualification stage (Stage 7), the stage at which the bidders are selected, based upon (among other things) their knowledge of a specialised sector of public services and their ability to manage risk.
   - General cost advice. This is the traditional quantity surveying role of advising on capital costs, including the preparation of preliminary estimates, bills of quantities, obtaining specialist quotes etc. In addition, value management and value engineering techniques (see Chapter 2) are increasingly being called upon to produce cost-effective design solutions.
   - Reviewing bids prior to submission – due diligence.
   - Advice on whole life costs. It has already been stated that to many the key to running a successful PPP contract is control of whole life costs. Recognising this, many surveying practices now have in-house advice available in this field.
   - Specialist advice. Obviously, highly complex projects such as the construction and management of a major hospital require a great deal of specialist input from, for example, medical planners, who are able to advise on medical equipment etc. from the outset. Surveying practices committed to developing their role in PPP already have at their disposal such expertise, which in some cases is in house.
2. *For funders*:
   - Due diligence. The financial and funding aspects of major projects are becoming increasingly susceptible to both technical and commercial risks. Investors and funding

institutions are looking more and more for independent scrutiny of all aspects of development, from design integrity to contractual robustness of the contract and beyond to the expenditure levels and progress against programme. The skills of the quantity surveyor provide an excellent platform for the investigative and analytical processes necessary to satisfy these requirements.

3. *For the public sector purchaser*:
   - Procurement advice. This method of procurement is, in many cases, just as unfamiliar for the public as for the private sector. The surveyor can advise the contracting authority on how to satisfy the requirements of this method of procurement. It is widely agreed that the appointment of a project manager at an early stage is vital to PPP project success. In addition, pressure is being exerted to speed up the procurement process – a factor that makes the role of the project manager even more crucial.
   - The outline business case. The preparation and development of the outline business case in Stage 3 involves the preparation of a risk register and the identification and quantification of risk, all of which are services that can be supplied by the quantity surveyor.
   - Advice on facilities management. Technical advice on this topic can be provided during the drawing up of the service specification, at Stage 3 and beyond.
4. *Common and joint services to SPCs/public sector purchasers – joint public/private monitor certifier*. This role is similar to the role played by a bureau de contrôle in France, and involves monitoring the construction work to ensure that it complies with contract. In addition to the built asset, the surveyor employed in this role can monitor facilities management operation. The concerns with this practice centre on the 'belt and braces' way in which the certification is being carried out, and the fact that firms are signing off multi-million pound schemes for very little fee and are effectively acting as unpaid insurance agents, with any claim being covered by professional indemnity insurance.
5. *For consortium building contractors*. This is the role recognised by many surveyors as the main involvement in PPP. It includes preliminary cost advice, preparation and pricing of bills of quantities, and supply chain management.

Emerging trends in PPP procurement

1. *Standardisation of contract conditions.* Many of the consortia now bidding for PPP/PFI projects are experienced in the procurement procedure to the extent that standard terms and conditions are increasingly being developed in order to speed up the procurement process.
2. *Separate competition.* Separate competition (i.e. outside of the main consortium) is emerging for:
   • Design – as previously mentioned, a pilot project is already under way in which the design of the built asset was separated from the rest of the deal
   • Finance – at least one PPP deal concerned with the maintenance of government offices has been concluded by finance secured by separate competition based on standard terms and conditions; the aims were to persuade banks and other project funders to accept standard contract terms for future PFI projects in the hope that this would greatly streamline the procurement process and reduce the time required by the due diligence procedures.
3. *Mixed sector projects.* The provision of services to housing, education and healthcare, for example, could be combined in a single deal, where the value of the individual deal is too small to make PPP possible.
4. *Monitoring the profit levels of PFI consortia.* How much profit should a consortium make? The Office of Government Commerce is currently looking at this question. Authorities and contractors have inherently different objectives, and there is a need to reconcile their different aims if their long-term relationships are to be successful. Only 15% of existing PFI deals have a provision that allows any windfall profits to be shared between the consortia and the public sector authority.

Conclusions

Potentially one of the biggest hurdles for PFI to clear in the UK is the opposition to the process by trades unions. From the outset there has been outspoken criticism of the PFI by unions, UNISON in particular, claiming that it is privatisation by the back door. There is a fear that PPP/PFI is just another step down the road of privatisation and marketisation, where the main criterion for service provision is complying with minimum legal standards, which are, the unions claim, below prevailing public sector levels. There is a list of high profile cases, including the London Underground, Portsmouth Hospital (where an Official Journal Announcement had to be withdrawn a week after publication) and Dudley Hospitals in the West Midlands, that are testaments to the power of the trades unions to halt a project dead in its tracks. In addition, 2001 saw over 20 PFI hospital projects put on hold because of lack of co-operation from public sector unions. To ease the situation and satisfy the trades unions, the General Secretariat of the Trades Union Congress has suggested restricting PFI to design, finance and build only, thereby leaving the employees in the public sector (as in the National Health Service). The private sector, on the other hand, would claim that this sort of restriction reduces the return on PFI projects to the point where it is marginal – the prison-building programme being a prime example of whole-scale transfer leading to large-scale savings and value for money. Already the Transfer of Undertakings (Protection of Employment) Regulations 1981, or TUPE Regulations, are in place, and these give unions a voice in the PFI process; however, much of TUPE applies only to existing employees, and new starters can be employed on different terms. Of course this means that during the time span of a PFI contract, all employees may eventually be employed on non-TUPE conditions. A more conciliatory approach towards conditions of employment for PPP projects has been taken by the Irish government, by engaging in so called 'stakeholder discussions' with the public sector workers as part of the procurement procedure. Finally, throughout this chapter the point has been made that PPP is now maturing; however, a major blow to the confidence of the financial backers of PPP projects was dealt when the government decided to withdraw its support from Railtrack in the autumn of 2001. In the long term the difficulties of this very peculiar privatisation scheme are expected to be minimal, but a short-term consequence was to send the bankers scurrying back to their bunkers, refusing to sign off contracts. Nevertheless, in what must be seen as the ultimate stamp of approval, in May 2000 HM Treasury concluded a PPP deal with a private consortium to refurbish and maintain its offices in Whitehall for the next 35 years. Whatever, the problems, PPP is a growing worldwide trend.

The international development of PPP

Today there is a growing acceptance around the world that public private partnerships have a place in the delivery of public services. Table 4.8 gives an indication of the extent to which PPP has been adopted in other parts of the world. Although this table is not comprehensive and is based on data current at May 2001, it clearly demonstrates that PPP is being examined and deployed in a large variety of sectors and countries throughout the world, with the UK leading the field by a long way in terms of the number of projects.

PPP in Europe

Surveyors and other PFI advisors are increasingly looking outside the UK for potential fee income as public private partnerships are increasingly being used in other European countries to carry out major capital projects, albeit without the same degree of commitment as in the UK. Two close European neighbours that are leading the field and are worthy of closer examination are Holland and Eire.

PPP in Holland

PPP is expected to become a significant aspect of government procurement in Holland, and it is planned to develop this once experience has been gained on the initial pilot projects. The Dutch government is trying to learn from the mistakes of the UK in pursuing too many projects without sufficient budgetary cover and, unlike in the UK, PPP enjoys widespread public support because of this cautious approach.

The stated goals of PPP in Holland are to achieve added value and improve efficiency. To date, the type of projects earmarked by the government for PPP would seem to fall principally into transport infrastructure (including roads and urban development). PPP in Holland is co-ordinated by The Dutch Ministry of Finance, which in January 1999 established The PPP Knowledge Centre (Kenniscentrum PPS) with the mission of initiating and promoting public/private co-operation. The procurement process is similar to the UK model, with promoters needing to demonstrate value for money by means of a public/private

Table 4.8 International development of PPP

	Number of deals	Roads	Rail	Water	Waste	Health	Prisons	Education	Power
Australia	> 20	+	+	+		+	+		
Belgium	> 10	+	++	+	+	+		++	
Canada	> 100	+	+	+	++				
Eire		+	++	+	++		+	++	+
Finland	> 5	+	++	+		+			
France	> 100	++	++	++	++				
Germany	> 5	++	++	++		++		++	
Greece	> 5	+	++	++	++				+
Holland		+	++	++					
Italy	> 5	++	+		+				
Japan	> 5				++				
Portugal	> 10	+	++	+		++			
S. Africa	> 10	++		++		++	+	++	
Spain	> 5	+	+			+			
Sweden	> 5		+						
UK	> 300	+	+	+	+	+	+	+	

+, deals closed; ++, deals proposed.

comparator before the commencement of the tendering process. A public sector comparator is then used to compare final bids against the alternative public sector contracting, based on a non-prescriptive output specification. Projects are thus far divided into two categories:

1. National projects, which are often in the form of a concession and are similar in nature to the UK PFI model, where a government agency pays for the delivery of a service. An example is the A59 Rosmalen to Geffen motorway, for which the design, construct, finance and maintain procurement procedure was started in 2000. As with the UK model, the following criteria must be demonstrated or met:
   • Rigorous risk analysis and allocation
   • Full business case
   • Payment mechanisms
   • Clearly defined output specification
   • Compliance with EU public procurement directives.
2. Combination projects, which are similar in operation to the UK PPP/4Ps model in that a proportion of the finance is supplied from public funds. Interestingly, Dutch Ministry of Finance guidance on this type of project specifically draws attention to the importance of financial benefit sharing on any future better-than-expected returns on investment. The Rotterdam Mainport Development Project is an example of a combination project.

An interesting point to note concerning the Dutch approach is that there are the means to reimburse a proportion of the bid costs to consortia, and this has already been done in several major projects to date. Also, Dutch planning law provides a mechanism for clawing back some of the enhanced value that accrues to a PPP project over its contract life.

PPP in Eire

PPP has been operating successfully in Eire for a number of years – for example, the East Link and West Link bridges in Dublin have been operating on a concession financially freestanding basis for a number of years. New impetus was given to PPP when the Irish Government established a dedicated PPP unit in 1999, and in its National Development

Plan 2000–2006 set the target for investment through PPPs at €2.35 billion. Over 500 projects were identified in the four sectors concerning roads, water, waste management and public transport (which includes the Dublin Light Railway Project). One of the main drivers for the adoption of PPP in Eire is the huge strain being put on the existing infrastructure by the doubling of the Irish economy during the period 1993 to 2000. The genesis of PPP in Eire has therefore been significantly different to that in the UK. The drivers in Eire are the accelerated delivery of national priority infrastructure projects, together with value for money over the full life cycle of the asset, rather than the procurement of public service assets off balance sheet. In the UK the initial projects involved NHS trusts, prisons and defence facilities, while in Eire the main emphasis has been on toll roads, public transport and the water supply and treatment sectors.

References

Audit Commission (2001). *Building for the Future, The Management of Procurement under the Private Finance Initiative.* HMSO.

Arthur Andersen and Enterprise LSE (2000) Value for Money Drivers in the Private Finance Initiative. HMSO.

Bates, M. (1997) Review of the Private Finance Initiative. HMSO.

HM Treasury (1998). *Policy Statement No. 2 – Public Sector Comparators and Value for Money.* HMSO.

HM Treasury (1999). *Technical Note No. 5 – How to Construct a Public Sector Comparator.* HMSO.

Institute for Public Policy Research (2001). *Management Paper, Building Better Partnerships.* Institute for Public Policy Research.

National Audit Office (1999). *Examining the Value for Money of Deals under the Private Finance Initiative.* HMSO.

National Audit Office (2001). *Managing the Relationship to Secure a Successful Partnership in PFI Projects.* HMSO.

PricewaterhouseCooper (1999). *PFI Competence Framework – Version 1.* HMSO.

Further reading

4Ps (2000). *Calculating the PFI Credit and Revenue Support for Local Authority PFI Schemes.* HMSO.

Buckley, C. (1996). Clarke and CBI unite to revive PFI. *The Times,* 22 October.

Building (2001). PFI plan to liberate architects, 31 August.

Catalyst Trust (2001). *A Response to the IPPR Commission on PPPs.* Central Books.

HM Treasury (1998). *Stability and Investment for the Long Term.* Economic and Fiscal Strategy Report – Cm 3978. HMSO.

Kelly, G. (2000). *The New Partnership Agenda.* The Institute of Public Policy Research.

Ministry of Finance (2000). *Public–Private Partnership – Pulling Together.* PPP Knowledge Centre.

National Audit Office (2001). *Innovation in PFI Financing, The Treasury Building Project.* HMSO.

PricewaterhouseCooper (2001). *Public Private Partnerships: A Clearer View.* HMSO.

Robinson, P. (2001). PPP tips the balance. Public Service Review PFI/PPP 2001. Public Service Communication Agency Ltd.

Robinson, P. *et al.* (2000). *The Private Finance Initiative – Saviour, Villain or Irrelevance?* Institute of Public Policy Research.

Rose, N. (2001). Challenges to procurement: the IPPR and Byatt reports. *Government Opportunities*, 24 Jul.

Thomas, R. (1996). Initiative fails the test of viability. *The Guardian*, 22 October.

Treasury Taskforce (1997). *Partnerships for Prosperity – The Private Finance Initiative.* HMSO.

Waites, C. (2001). Are we really getting value for money? *Public Service Review PFI/PPP 2001.* Public Service Communication Agency Ltd.

Web sites

www.audit.commission.gov.uk
www.detr.gov.uk
www.dfee.gov.uk
www.doh.gov.uk
www.hm-treasury.gov.uk
www.minfin.nl/pps
www.nao.gov.uk
www.mod.gov.uk
www.partnershipuk.org.uk
www.pfi-online.com
www.ppp.gov.ie
www.4Ps.co.uk
www.unison.org.uk

5

Procurement – doing deals electronically

Introduction

This chapter continues with the theme of doing deals, and examines the changing environment and nature of procurement in general and the impact of electronic commerce on quantity surveying practice in particular.

e-Commerce/e-construction

Since the emergence of the term *e-commerce* into the wider public domain in 1997, it has seldom been out of the media spotlight. After the meteoric rise in the value of the so-called dot.com companies during the late 1990s there came the inevitable crash, with reports of thousands of small investors losing millions of pounds when share prices collapsed during 2000–2001. This series of events seemed to confirm what many construction industry pundits had been voicing publicly about e-commerce; that it was merely a 5-minute wonder that enabled an elite number of business analysts and bankers to cream off vast profits, and a small percentage of the population to buy books or theatre tickets on-line. The general consensus is that the failure of many dot.com companies occurred owing to:

*   Poor business models
*   Poor management
*   Aggressive spending
*   The companies forgetting the customer
*   The changing attitudes of venture capitalists.

However, e-commerce is much more than ordering weekly groceries from a web site and having them delivered to your door; e-commerce and the digital economy are bringing about a revolution in business practices and, despite the birthing pains, are here to stay. The digital economy cannot and should not be ignored, as it will have a major impact on working practices for the quantity surveyor and the construction industry, thereby making a significant contribution to delivering added value. In common with other market sectors, e-construction has had a number of false dawns. For example, at the height of dot.com mania, five of the largest contractors in the UK announced the creation of the first industry-wide electronic marketplace offering the purchase of building materials online as well as a project collaboration package. Just over a year later it was announced that the planned Internet portal was to be shelved owing to lack of interest. Despite this lack of enthusiasm from within the industry and the professions, the move to become an e-enabled industry continues – due in large part to the continuing client-led drive for improvement, as well as successful examples from other sectors. Perhaps a note of realism has also been sounded, as contractors now seem to understand that their core business is not suited to running these kinds of ventures, and as a result they are turning to specialist providers to supply the technology and skills required. There is also the awareness that companies cannot provide every kind of service. Consequently, the future for e-commerce seems to lie, at least within the property sector, with niche market provision. One thing that all sectors of e-commerce are sure about is that the more fragmented the market, the more efficiency benefits e-commerce ventures can bring, by uniting the disparate elements of the supply chain. Information technology is at the heart of the developing tools and technologies that pool information into databases. An equally important aspect is looking at the attitudes of the people who need to feed information into and use the system, and this will be discussed in detail in Chapter 6. The Internet in particular provides a platform for changing relationships between clients, surveyors, contractors and suppliers; open exchange of information is critical in order to harness the best from this virtual marketplace, and one of the biggest challenges is creating a culture that encourages and rewards the sharing of information. Too many people still have the viewpoint that

knowledge is power and commercial advantage, and the belief
that the more that they keep the knowledge to themselves, the
more they will be protecting their power and position. The transparency of the Internet should be a driving force for changing
business strategies and attitudes, and yet it will take a quantum
leap in construction business culture before sensitive information is disclosed to the supply chain. It has been suggested, by
a leading construction industry dot.com, that the European
construction industry could save up to £120 billion per annum
on building costs and reduce completion time by up to 15%
through the adoption of e-construction technologies. Until
recently, business on the Internet was dominated by technology-
driven companies selling well-financed ideas. These start-up
dot.com companies enjoyed the luxury of abundant investment
capital without the burden of having to show a profit, but that
time has come to an end. Although initially they lagged behind
the 'idea companies', traditional 'bricks and mortar' companies
are quietly making up for lost time and going online. The
current wisdom is that it will probably be these companies that
will ultimately become the financial cornerstones of e-commerce.

When lean thinking first hit construction, it was the car
industry that provided the role model (see Chapters 1 and 4).
Now that some sections of the construction industry are
seriously talking about e-commerce, they can once again look to
the motor industry for a lead. In America, the three major
domestic motor manufacturing firms have been dealing with
suppliers via a single e-commerce site for a number of years –
an initiative that has resulted in reported savings of over £600
million a year. Similar initiatives are also to be found in the
retail and agriculture sectors, but perhaps in the rush to establish the first truly successful construction-based e-portal, the UK
players ignored some basic business rules (Figure 5.1).

According to the European Commission Electronic Commerce
Team's report *Just Numbers* (Hobley, 2001), 70% of all European
small to medium-sized enterprises (SMEs) currently have access
to the Internet and 40% have their own web site. Of the enterprises with a web site, the majority can be termed a first-gener-
ation presence – that is, the web site is used for marketing
purposes, using the simplest of business models. The so-called
second and third-generation presence, which incorporates transaction applications, shows a much slower growth rate. So why
are companies using e-commerce?

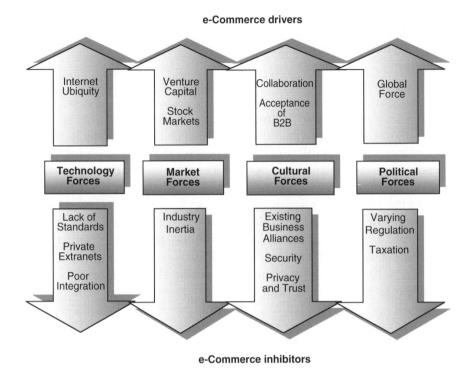

Figure 5.1 e-Commerce drivers and inhibitors (Source: B. Mathieu, Gartner Group).

Companies with Internet access identified the following as the main factors for Internet use (source: Deloitte Consulting):

Information on markets/competitors	79%
Promotion	51%
Distant collaboration	46%
After sales services	32%
Business to business (B2B)	27%
Business to consumer (B2C)	26%
Recruitment	18%
Response to public tender	13%

In addition, the benefits that SMEs perceive that they derive from the use of the Internet are:

Knowledge of competitors	67%
Facilitate partnerships	55%

Faster reaction	48%
Lower communication costs	48%
Expansion into new markets	42%
Simplify administration	25%

The figures shown above indicate that, to date, the majority of organisations would seem to utilise e-commerce to track the competition and improve communications. In addition it would seem that many companies, particularly SMEs, are engaging in e-commerce activities as a result of competitive pressure, suggesting a defensive line of action rather than a differentiated one. However, as clients become more e-enabled, quantity surveying practices must follow or be left behind. As in the case of supply chain management techniques (see Chapter 2), the pressure will come from the client; even so, there are as yet very few surveying practices that have made the leap from simple web site to transaction platforms.

e-Commerce defined

e-Commerce exploits information and communication technologies (ICTs) to re-engineer processes along an organisation's value chain in order to lower costs, improve efficiency and productivity, shorten lead-in times and provide better customer service. It may therefore be defined as:

> Any net-enabled business activity that transforms internal and external relationships to create value and exploit market opportunities, driven by the new rules of the connected economy.
>
> (Gartner Group)

For the quantity surveyor, e-commerce allows instant communication through the supply chain, giving the partners a clear real-time picture of supply and demand. One of the first organisations to use the term e-commerce was IBM, when in October 1997 it launched a thematic campaign built around it. Today major corporations are rethinking their business in terms of the Internet and its new culture and capabilities, and construction and surveying must follow this trend. As with any new innovation, there are forces that act to drive forward the new ideas as well as forces that act as inhibitors to progress, and these are summarised in Figure 5.1.

e-Commerce can be said to be the ultimate supply chain communication tool, as it permits real-time communication between the members. e-Commerce can include the use of some or all of the following technologies:

- *The Internet* (the international network). This is the publicly and generally freely accessible information superhighway. Initially, within Europe a major barrier to the take up of Internet use was the comparatively expensive telecommunications costs of monopoly-owned telecommunication services such as British Telecom, etc. However, EU legislation has liberated the market to the point where today there are nearly 900 telecommunication enterprises in Europe, which has resulted in a consequent decrease in call charges. The main advantages of the Internet include availability, low cost and easy access, whereas the main disadvantages (particularly for business users) centre on the lack of control, reliability and security – aspects that are now being addressed and will be discussed later in this chapter. For the surveyor, Internet applications include procurement, marketing, e-mail and data transfer. For example, a significant number of hard-pressed UK quantity surveying practices transfer project drawings, in CAD format, via the Internet to practices worldwide for the measurement and preparation of bills of quantities. The completed bills plus drawings are then e-mailed back to the UK, permitting virtually 24-hour working.
- *Intranets.* These are internal networks that publish, for example, information available to staff within a single company – not the world. Compared with the Internet, intranet sites are much faster to access and offer great savings in set-up, training, management and administration. An intranet is a very cost-effective way of centralising information sources and company data, such as phone lists, project numbers, drawing registers and quality procedures, as well as allowing the use of internal e-mail. Intranets use the same technologies as the Internet, but are not open to public access.
- *Extranets.* Extranets are wide area intranets that span an organisation's boundaries, electronically linking geographically distributed customers, suppliers and partners in a

controlled manner – i.e. it is a closed electronic commerce community, extending a company's intranet to outside the corporation. It enables the organisation to take advantage of existing methods of electronic transaction, such as Electronic Data Interchange (EDI), a system that facilitates the transmission of large volumes of highly structured data. Project extranets have been described as the first wave of the e-commerce revolution for the UK construction industry, and applications that utilise extranets include project management (for example, the construction of Hong Kong's new airport). For some, EDI promised the ability to exchange data efficiently between trading partners, as had been the case in the motor industry and food retailers for a number of years. The major disadvantages of EDI are the high cost, as operators have to trade through value added networks (VANs), and the standards problems such as incompatibility. More importantly, the point-to-point contact of EDI provides no community of market transparency. These problems are increasingly being addressed by the reduced costs of Internet applications that are able to deliver flexibility, reduced training costs and low set-up costs. An example of the limitations of EDI is provided by the experiences of H&R Johnson Tiles, the largest manufacture of tiles in the UK. This company has carried out electronic trading for years with its larger customers; however, they have found it impossible to extend EDI to smaller customers without the critical mass of transactions to drive the necessary investment; and have had to launch a separate web site using an extranet to cater for its top 20 smaller customers. This is not to say that batch-mode EDI transactions will not survive and prosper, as the system is extremely efficient and has been predicted to have an established place in the large-scale exchange of data.

Little wonder, then, that with the pressure to do business electronically the process of choosing the right solution can be somewhat confusing, leading to decision inertia among some sections of the professions and industry. However, despite the claims and counterclaims there is one undeniable fact about electronic commerce: the revenues it generates are truly immense and are predicted to rise, in all sectors, to US$1533 billion in Europe and US$6.8 trillion in the USA by 2004.

e-Commerce in the UK construction and property industries alone has been predicted to rise to £67 billion, and £4000 billion globally, by 2004. The question for many construction related services, including surveying, is, will e-commerce transform business practice from 'bricks and mortar' to 'clicks and mortar'? The surveying practice thinking of launching into e-commerce must view the venture holistically – e-commerce is not just the introduction of cutting-edge technology, but the integration of technology into existing business plans to introduce new working practices. When drawing up a business plan that includes an e-commerce application, one of the fundamental questions for a surveying practice must be, what capital and long-term costs are involved, and what return can be expected? When questioned about the perceived obstacles to adopting e-commerce, organisations cited the following:

- *Current e-climate.* The volatility of dot.com business has given rise to much publicity and discussion. Attrition rates are much higher among e-business dot.coms than among SMEs in more traditional businesses; a report by the Jupiter Investment Bank in February 2001 forecast fewer than 100 of today's 500 Internet marketplaces will survive. This may act as a further disincentive to SMEs, as bankruptcy laws act as a strong disincentive to risk taking, particularly in the UK.
- *Regulation.* Under e-commerce law, the regulation and security aspects of a virtual trading environment cause many enterprises concern. In a survey carried out among European SMEs by eMarketer Inc. in 2001, 47% of companies surveyed sited lack of legal guarantees and trust for online transactions as a major issue.
- *Skills.* There is currently (November 2001) estimated to be a shortage of 1–1.5 million IT specialists in the European Union; little wonder that lack of skilled staff plus long training periods for existing staff is a concern! An initiative, 'Go Digital', was launched by the European Commission in April 2001 to encourage SMEs to take up IT and e-commerce.
- *Technology and standards.* Concerns were raised regarding which of the alternative standards to adopt and which technology to invest in – and which systems would be proof against costly upgrades in a few years time.

- *Costs*. Levels of investment and involvement in e-commerce can vary considerably according to the size and nature of the organisation. Level 1 is the entry level, is limited to simple use and requires minimal investment in hardware, software and human resources – a perfect platform for the small practice. Level 3 is a virtual organisation that requires a major investment in human resources as well as industry-wide culture and process changes. As indicated in Table 5.1, the costs associated with entry level to e-commerce are minimal – probably less than a season ticket to Arsenal or Manchester United – and therefore the perception that involvement attracts a high cost is a false one. Nevertheless, questions that should be addressed at the outset include: at what level (both in terms of cost and commitment) should an organisation enter e-commerce? And which business model should be adopted?

e-Commerce markets

e-Commerce can be broken down into categories, based on who is trading with whom. Latterly, sectors such as construction, finance etc. have coined their own terms – e.g. e-construction, e-finance etc. – to stake their own unique claim in the electronic marketplace. Although some sources claim that there can be up to nine classifications of e-commerce, most people agree that there are only three:

1. Business to business (B2B)
2. Business to consumer (B2C)
3. Consumer to consumer (C2C).

Of these, only the first two have seen dramatic growth:

- The business-to-business (or business to administration, B2A) sector includes aspects such as online data exchange, and usage is predicted to rise to 46% of all businesses by 2002. For the quantity surveyor it includes procurement online and access to real-time databases, and usage is projected to rise to 49% of all businesses by 2002. This category of e-business utilises the Internet and extranets, and it is forecast that spending in B2B is expected to dominate Internet growth until 2003. There are basically two different types of B2B companies; horizontal and vertical. As illustrated in Figure 5.2, vertical B2B companies work within an industry and typically make money from advertising on specialised sector-specific sites or from transaction fees from the e-commerce that they may host (e.g. BuildOnline.com). Horizontal B2B companies are a completely different breed, and operate at different levels across numerous different verticals. Whether it's enabling companies electronically to procure goods, helping to make manufacturing processes run more efficiently, or empowering sales forces with critical information, most horizontal companies make their money by selling software and related services (e.g. www.Tendersdirect.co.uk).

- The business-to-consumer (B2C) sector includes financial services and banking online, and is predicted to rise to include nearly 24 million users by 2005. For example, the financial sector has been an enthusiastic adopter of B2C, as bank transaction costs can be reduced by up to 95% with Internet-based services. The travel and accommodation sector, entertainment, and shopping online (including, books, CDs and videos etc.) have followed suit, and usage has been

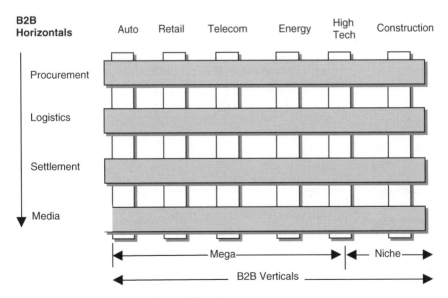

Figure 5.2 Structure of B2B business
(Source: PricewaterhouseCooper).

predicted to rise to 61% of EU population by 2003. The motive is simple: added value. It has been estimated that issuing an airline ticket manually costs £5, whereas it costs just £0.52 via the Internet. This sector mainly utilises the Internet, but despite the attractions for companies, B2C is considered a highly unstable sector that employs, in some cases, questionable accounting practices.

The development of B2B commerce has been breathtaking when compared with other technological innovations. Figure 5.3 illustrates the development of B2B commerce; after the establishment of closed EDI networks, the second phase of e-commerce saw the emergence of one-to-one selling from web

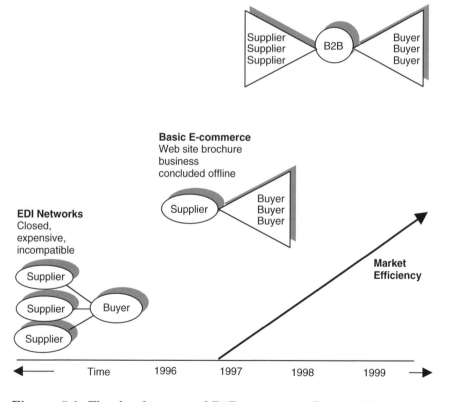

Figure 5.3. The development of B2B commerce (Source: Morgan Stanley Dean Witter Internet Research).

sites and a few early adopters began pushing their web sites as a primary sales channel (e.g. Cisco and Dell). Most of the initial web sites were not able to process orders or to supply order tracking. The current phase of development is represented by the butterfly model, and involves the rise of third-party sites that bring together trading partners in a common community. In this situation buyers and sellers start to visit the site regularly, thus moving to the collaborative phase, which creates the opportunity to serve a large percentage of those interests. In a few years B2B has developed from a high-cost rigid system with low transparency to a low-cost highly flexible and fully transparent system.

The development of e-commerce towards e-business (the more comprehensive and preferred definition) reveals the move from the enterprise-centric vision to the multi-enterprise network, or virtual enterprise, and the move towards the exploitation of the ICT and the potential of the Internet from cost reduction-oriented e-commerce to the collaborative-commerce vision – a move that requires a major step change in business culture. The challenge for construction and the surveying professions is undoubtedly greater than for some areas – Dell, for example, has the considerable advantage of being in the Internet-related product field, where clients and suppliers are already technologically sophisticated – but nevertheless, real value added benefits are available for the quantity surveyor. The key point now currently acknowledged is that, with the usage of ICTs and the Internet in business, not only can costs be reduced, but value can also be created. Value is created from brokering transactions and matching orders between companies, but also from the provision of additional services such as professional services for integrating and managing companies – including legal and financial services, logistics, project and contract management, as well as background services such as market intelligence. Dell, for example, has created virtual integration with both its upstream partners and its downstream clients so that the entire supply chain acts like a single integrated company. Dell builds computers to order; typically, someone who works for a large company like Boeing goes to a private web page available only to Boeing employees and can order and configure a computer online. Dell suppliers maintain a 2-week supply of components near to Dell factories; this inventory belongs to the suppliers and not Dell. Dell shares information with suppliers regarding inventory levels, sales and forecasts, and works with suppliers as a virtual enterprise.

Application of e-business to surveying/construction

One of the most discussed topics in e-commerce is that of business models because, just like conventional business, e-business needs to make a profit. As many companies like Boo.com have found to their cost, a key component in making a profit is a robust business model. Today this is more important than ever as the cold wind of reality hits e-commerce. Gone are the days when venture capitalists were falling over themselves to find e-commerce ventures to back. What is needed now is a proven business case; funders want to see profits before they'll hand out any more cash. When a potential investor in an e-commerce project asks the question 'what is your business model?', they are really trying to discover where the business is going to make money and why people going to pay for using that particular service. A business model should give product/service information and income generation flows and, together with the marketing strategies, enable the commercial viability of the project to be assessed. An e-business model makes it possible to answer questions such as: How is competitive advantage being built? What is the positioning and the marketing mix? In theory very many new business models can be conceived; however, in practice only a limited number are being realised in electronic commerce. See figure 5.10 for an illustration of 11 business models, which are mapped along two dimensions to indicate:

- The degree of innovation, which ranges from essentially an electronic way of carrying out traditional business (in the bottom left-hand corner) to value chain integration (in the top right-hand corner), a process that cannot be done at all in a traditional form as it is critically dependent on information technology to let information flow across networks
- The extent to which functions are integrated.

The models that have applications in quantity surveying/construction are as follows:

- *e-Shop – seeking demand.* This can be the web marketing of a company to promote, in the first instance, the services that it provides. Seen as a low-cost route to global presence, it is now almost obligatory for quantity surveying practices to have their own web site. Its function is primarily promotion, although it has to be sought by prospective clients.

- *e-Mall – industry sector marketplace.* An electronic mall in its basic form is a collection of web sites, usually under a common umbrella (e.g. www.propertymall.com). The e-mall operator may not have an interest in the individual sector and income is generated in return for fees paid by the hosted web sites, the fees usually being composed of an initial set-up fee plus an annual fee. The cost of this sort of platform package can be comparatively modest, depending on the size and the complexity of the web site. In addition, some marketplace portals permit related organisations (say, a quantity survey-ing practice) to list limited company details for no charge, either permanently or as an introductory offer. Construction has large volumes of buyers and sellers and geographically dispersed sites, and remote buyers and sellers may be brought together, even when based in different countries. An online marketplace cuts the cost of communication, reduces commu-nication errors, and speeds up the entire procurement process.

- *e-Procurement – seeking suppliers.* This is the electronic tendering and procurement of goods and services. Benefits are said to include a wider choice of suppliers (leading to lower cost), better quality, improved delivery, and reduced cost of procurement. Electronic negotiation and contracting and possibly collaborative work in specification can further enhance time and cost savings and convenience. For suppliers the benefits include more tendering opportunities (possibly on a global scale), the lower costs of submitting a tender, and possibly tendering in parts that may be better suited for smaller enterprises or collaborative tendering. Lower costs can be achieved through increased efficiency, and the time may not be too far distant when all procurement is done this way.

- *e-Auction.* This offers an electronic implementation of the bidding mechanism used in traditional auctions. The system may incorporate integration of the bidding process together with contracting and payment. The sources of income for the auction provider are from selling the technology platform, transaction fees, and advertising. Benefits for suppliers and

buyers include increased efficiency and time savings, the fact
that there is no need for physical transport until the deal
has been established, and global sourcing.

- *Collaboration platforms.* These provide a set of tools and an
information environment for collaboration between organisations. This can focus on specific functions, such as collaborative design and engineering, or providing project support
with a virtual team of consultants. The business opportunity
is in managing the platform and selling the specialist tools
(e.g. www.BuildOnline.com).
- *Value chain service providers.* These organisations specialise
in a particular and specific function of the value chain – for
example, electronic payment or logistics. A fee or percentage-
based scheme is the basis for revenue generation.
- *Value chain integrators.* These focus on integrating multiple
steps in the value chain, with the potential to exploit the
information flow between these steps as further added value.
- *Virtual communities.* Perhaps the most famous of virtual
communities is the ubiquitous Amazon.com.

Information brokerage, trust and other services

Trust in e-commerce

As mentioned earlier, trust or, to be more accurate, the lack of
trust in Internet commerce is a major concern for most organisations. Widespread adoption of e-commerce depends on users
having trust and confidence in the whole activity.

Identity theft is a major and growing problem in conducting
transactions over the Internet. Identity theft occurs when a
person or organisation poses as someone else in order falsely to
pass payments for goods or obtain digital goods, create false
purchase orders, falsely trade on an organisation's credentials,
or solicit personal information. Within the business-to-business
environment, rigorous identity and knowledge of who is being
dealt with is essential, especially when business partners may
be spread around the world. There must be confidence that
electronic messages have not been tampered with and that their
contents have been kept confidential. Businesses trading on the
web need to ensure they can distinguish between legitimate
clients and fraudulent users in real time, and have the assur-

ance that the acceptance of contract conditions will be honoured and not subject to fraud. Users need to be reassured about privacy, and about how their personal and payment information will be protected and not abused.

Internet connections will drive the demand for more secure online services. According to IDC, the number of Internet users worldwide will grow from an estimated 69 million at the end of 1997 to an estimated 320 million by 2002. Growth could therefore exceed 400% in just 3 years. However, such growth rates are critically dependent upon both consumers and businesses having trust in undertaking transactions online.

However, the media reports, almost on a daily basis, how trust in the Internet has been seriously compromised both in the UK and abroad. Governments, consumer associations and industry groups around the world have identified the lack of trust as the major factor that is holding back the development of e-commerce. According to research undertaken for the Department of Trade and Industry in the UK, 60% of Internet companies have reported a security breach during the last 2 years. In August 2000, Barclays Bank plc reported that failures in its Internet banking system enabled online banking customers to access the wrong bank accounts. According to Experian, credit card fraud is now as high as 40% for some UK online retailers. The impact of not having trust can be devastating, and leads to low conversion rates of visitors to buyers on web sites. Most consumers do not trust site owners to the extent that, according to Market Explorers (US), 37% of site visitors give false information.

- Only 3% of users are always comfortable about entering transactions online (BBC Breakfast News)
- 60% of Internet users who do not shop online stated privacy and security concerns as the main hurdle (Privacy Council, as reported on Reuters)
- 50% of online shoppers are concerned about privacy and security (SWR Worldwide)
- 20% of websites have detected unauthorized access (FBI).

Online fraud is growing, and the size of the problem is alarming:

- Fraudulent credit card activity is 12 times more likely online than offline (Gartner Interactive)

- Fear of fraud is the number one reason for users deciding against making online purchases (Web Assured Survey)
- 64% of online consumers are less likely to trust a web site, even with a privacy policy (Jupiter Communications)
- 25% of online orders are not fulfilled properly
- More than 92% of reclaims of transactions from vendors by credit card companies ('charge-backs') come from Internet transactions (Gartner Group)
- According to a Canadian survey, over 80% of users are concerned about privacy online (JC Williams Group Ltd, as reported in Reuters)
- Online fraud could reach $60 billion by 2005 (Meridian Research)
- Visa Canada reports that more than 50% of disputed transactions originate on the web, although only 1–2% of Visa transactions are processed online
- User name and passwords offer inadequate protection.

Identity theft, or someone impersonating someone else or a business, is a serious issue. Cable & Wireless recently mounted an advertising campaign warning of fraudulent purchase orders being issued (Figure 5.4).

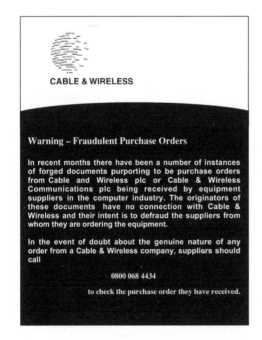

Figure 5.4 Cable & Wireless advertisement.

According to Jodie Bernstein, Director of the Federal Trade Commission's Bureau of Consumer Protection, the fear of identity theft has gripped the public as few consumer issues have.

The tools and information are easily available. Hacker sites on the Internet publish programs that help the criminal commit fraud. There are programs that can crack user name and password access to web sites and systems, and credit card generators that can create thousands of valid credit cards in seconds (which can then be used on web sites). Similarly, Social Security numbers and address information can be quickly obtained to create false identities for citizens of different countries.

Trust infrastructures are vital if e-business is to develop a secure trading environment. Without trust, e-commerce will fail. Trust is a fundamental hygiene factor in whatever environment. Trust is intangible but nonetheless central to all e-business success, whether business-to-consumer or business-to-business. It cannot be bought out of a box, and it encompasses (through a combination of technology, services and businesses processes) security, privacy, identity and customer service.

Businesses have increasingly looked for solutions that provide trust between parties. Digital certificates are the emerging critical tools that guarantee the user's identity. Typically, a digital certificate is only issued after completion of various checks designed to verify the user's identity. Typical verification sources include passports and electoral rolls for individuals, and Companies House and alternative corporate registration documents for organisations. The identity checks made before the certificate is issued are such that substantial liability cover can be offered. So if digital certificates are such a great idea, why are they not widely available? This is primarily down to the lack of availability of relevant applications that drive their adoption. However, one application where they are being rapidly introduced is in the area of electronic tendering.

Digital certificates and electronic tendering

Tendering is big business. As discussed in Chapter 7, in the European Union, public sector construction-related procurement alone is worth around 200 billion euros per annum. It's a critical procurement process, and for the public sector much is

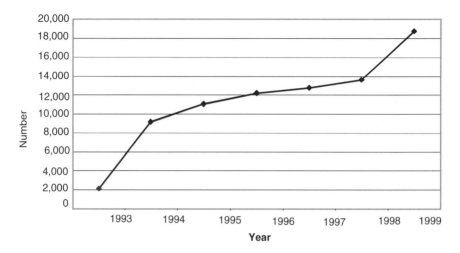

Figure 5.5 EU Total tendering entities.

regulated by law. Tenders above certain thresholds must be published in the *Official Journal of the European Communities* (OJEC), and the number of entities in the tendering process is increasing (Figure 5.5). Until now tendering has been very much a paper driven process, both time consuming and inefficient, and too often it has been the companies with sufficient resources and the administrative infrastructure that respond to them. This is a market that is ripe for electronic commerce, and where trust in the transaction is an absolute necessity.

Electronic tendering enables the traditional process not only to be made more efficient but also to add significant value. It can provide a transparent and paperless process that allows offers to be more easily compared according to specific criteria. More importantly, by using the Internet, tendering opportunities are available to a global market.

The e-tendering process

With any form of e-commerce, trust, security and confidentiality are paramount. With electronic tendering, where documents travel across the Internet – an inherently insecure medium – this is absolutely crucial. Purchasers and suppliers must have confidence in the electronic tendering process if it is going to be adopted and its benefits fully realised.

TenderTrust® is the world's first smartcard-based electronic tendering system that covers the entire tendering cycle. Trust is addressed by the use of x509 banking strength digital certificates provided by The Royal Bank of Scotland's e-Trust Services. TenderTrust meets the highest security and authentication standards, to enable suppliers and purchasers to manage the entire tendering process easily. Digital certificates are used not only to confirm the identity of the sender but also to sign and encrypt documents digitally so that they can be sent in a highly secure manner. Both purchasers and suppliers use smartcards to access the system (Figure 5.6).

The purchaser's smartcard enables only authorised persons to create and publish the tender documentation on the system, and it also enables them to open the tender responses once the documents are released after the closing date. The supplier's smartcard enables authorised personnel to retrieve the full tender documentation and to submit their responses securely. Suppliers can have total confidence that their bids are completely secure, signed and sealed, until the tender closing date.

A key benefit associated with the use of smartcards is that they provide an irrefutable audit trail, underscoring the transparent nature of the process. Every access to the system, and every document movement between purchasers and suppliers, is tracked with the automatic generation of receipts at critical points in the process. In addition to the downloading and uploading of documents, the smartcard is used to manage access to online supplier forums. Only those directly involved in a specific tender can post and view replies clarifying the tender.

Electronic tendering using smartcard technology offers substantial benefits for both purchasers and suppliers. For

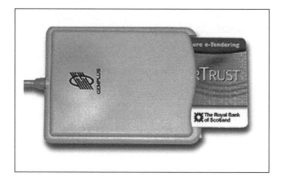

Figure 5.6
e-Procurement
smartcard.

purchasers, it means tenders can be downloaded securely at the touch of a button, enabling a shorter and more efficient tendering process. Other benefits include electronic publication of opportunities to an increased supplier base, easier supplier selection and qualification, improved supplier communication, and faster availability of tender documentation. Substantial cost and time savings can be obtained by elimination of the traditional paper chase, enabling greater concentration on areas where real value can be delivered – such as tender evaluation. In addition there is an irrefutable audit trail of the entire tendering process, from submission to award.

Suppliers are not only able to receive relevant invitations to tender as soon as they are uploaded by the purchaser; they can also send back their proposals securely and confidentially, safe in the knowledge that their bids can only be viewed at the tender closing date. They can also make changes and update proposals right up to the closing date. Trust is assured.

How TenderTrust® works

Step 1 (Figure 5.7):
- The purchaser prepares the tender documentation using the TenderTrust software. Typically this will include a Notice (summary of the tender), plus associated documentation.
- The purchaser uses a smartcard to log onto the system, and the digital certificate is checked to ensure that it is valid and

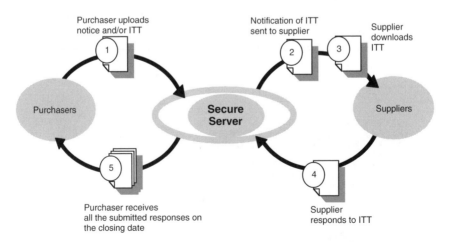

Figure 5.7. How TenderTrust® works

that the purchaser has the right to publish tenders on the system.

- The tender documents are then sealed, electronically signed, encrypted and sent to the TenderTrust server. Arrival at the server is time stamped, and e-mail notification of receipt is sent back to the purchaser.

Step 2:

- At the TenderTrust server the Notice is published on the public Internet site. The full document set is published to the secure server (only suppliers with a valid smartcard can retrieve the full documentation).
- Those suppliers who have set up a Tender Alert on the system and whose section criteria match that of the Notice are automatically notified by e-mail that there is a new tender opportunity.

Step 3:

- Access to the full tender documentation is available to suppliers using their smartcards. Before the documents are downloaded, the digital certificate on the smartcard is validated.

Step 4:

- The supplier prepares a Tender Response, attaching the appropriate documents. Different document types can be used – word processing, CAD, spreadsheet etc.
- The supplier logs on to the system, and the validity of the digital certificate is checked. The package of documents is sealed, electronically signed, encrypted and sent across the Internet to the TenderTrust server. Arrival at the server is time stamped, and e-mail notification of receipt is sent back to the supplier.
- The responses are held securely until the closing date, and these cannot be accessed.

Step 5:

- At the closing date, the purchaser, using a smartcard, can retrieve the tender responses from the suppliers for analysis and award. Again, the digital certificate is validated.

Digital certificate trust and security benefits

- Provision of controlled access to the e-tendering system for those who have a valid certificate
- Confidentiality – tender documents cannot be read by an unauthorised party

- Non-repudiation – neither party can deny having sent or having received tender documentation
- Integrity – no alterations can be made to the documentation
- Authentication of identity – to register, purchasers and suppliers have to provide detailed information to confirm their identity.

Key purchaser benefits

- Electronic publication of opportunities to an increased supplier base
- Tender documentation can be worked upon right until the last minute, and improved preparation time will increase the likelihood of quality responses
- Invitations to tender at the touch of a button; these tenders will be accessible for all eligible suppliers to upload as soon as the purchaser agrees the issue date and time
- Suppliers' tender responses are automatically received at the tender closing date
- Reduction in the paper chase and process
- Increased time for concentration on value adding activities such as tender analysis, rather than spending it on administrative tasks such as copying, collating, binding and distribution of tender information
- A complete and irrefutable audit trail of all document movements to and from suppliers; an instant record is made of when documents are sent and received, with associated receipts
- Improved management information about tendering activity and process
- Faster and easier communication with suppliers
- An integrated rather than a fragmented tendering process
- Faster pre-qualification of suppliers, ranked on predetermined criteria
- Shorter tendering timescales
- Cost savings in a more efficient tendering process, plus better value as a result of increased competition from the supplier base.

Key supplier benefits

- Suppliers can register their profile so that they are notified of ITTs' Invitations to Tender specifically relevant to their business

- A comprehensive search of all tender opportunities is published; this reduces costs and minimises missed opportunities
- Suppliers can make changes, update proposals and provide further information right up to the tender closing date, ensuring that the very best proposal possible is submitted
- Responses can be sent back at the touch of a button, increasing the amount of time spent on proposals rather than on the administrative process of printing, collating, binding and distributing documents
- Reduction in the paper chase, with no need to provide multiple copies of tender documents
- Suppliers are able to concentrate their efforts on value adding activities that could result in winning business, rather than spending time copying, collating and binding tender documents
- Easier communication with the purchaser through an electronic forum that facilitates amendments, commonly asked questions and answers
- Suppliers have an instant record of when all documents have been sent and received, and confirmation of receipt of documents is also issued; this guarantees a record and assurance of document transmission.

Cost savings

Comparing e-tendering and traditional tendering (Figure 5.8), market studies have shown that savings from electronic tendering systems arise in two main areas:

1. Improvement in the efficiency of the tendering process
2. Better bids leading to reduced procurement costs.

Merx, a Canadian e-tendering notification system (not smart-card based) has reported savings of around 15%, just with the improvements on tendering publication and notification. Savings can be achieved in the following areas:

- *Time and cost of tendering.* A shorter tender preparation cycle, the reduction in paperwork and improved communication (no couriers) can give savings of around 15–20%.

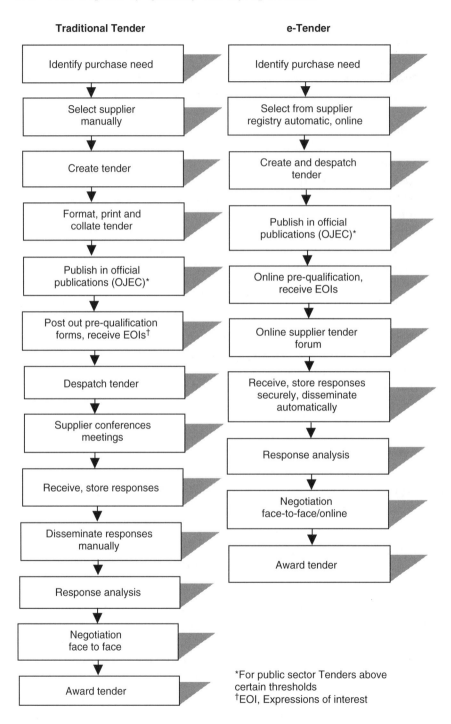

Figure 5.8 e-Tendering versus traditional tendering.

- *Improved quality of the tender issued.* Less time spent on the administrative processes involved in printing and sending tenders means that more time can be spent on the quality of the tender being issued. A better tender reduces the need for management to spend time on supplier meetings to clarify bids, and on telephone calls to check on tender progress. This can provide potential savings in management time of up to 20%.
- *Faster analysis.* Electronic analysis can perform pre-screening, scoring and rating of tender responses, to filter out those that are uncompetitive. Management time can be focused on improved evaluation and more detailed negotiation to get the best value from suppliers.
- *Improved competition.* e-Tendering increases the pool of supplier responses, since it removes one of the main barriers to participation – finding out about the tender. The costs saving through increased competition is estimated at around 15%.
- *Better quality of tender response.* Assistance can be provided through the use of wizards and proposal automation tools to improve the quality of the supplier response. This leads to savings in purchaser's management time in analysis.

The Office of Government Commerce's e-tendering pilot

The government is one of the major issuers of tenders. Traditionally government departments submit Notices and other tendering documents to publications such as the OJEC. Suppliers then follow a lengthy process of searching these sources to find appropriate matches for the services they offer. Once matches have been made, the suppliers have to request the full document set from the purchaser. This is usually made up of multiple lengthy sections that need to be circulated to various areas of the organisation. Preparing a tender response is a detailed process, and requires extensive input from many business areas to produce the various sections of the comprehensive document. Once this is complete, it has to be formatted, printed, collated and delivered on time. Once securely transported and received by the government departments, the multiple copies of the responses have to be stored until the closing date. The responses then have to be physically distributed to all those within the departments involved in the selection process.

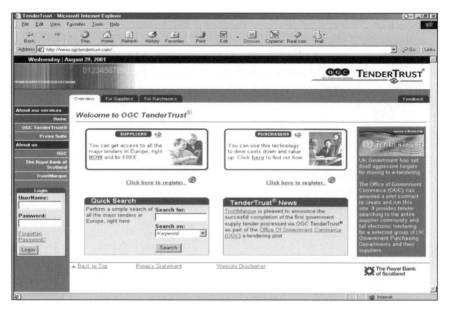

Figure 5.9 e-Tendering web page.

In June 2001, the OGC started a pilot scheme using TenderTrust to automate the entire process and deliver additional benefits and savings (www.ogctendertrust.com; Figure 5.9). The pilot scheme involves 10 leading government departments and a number of their suppliers in testing and evaluating the perceived benefits of OGC TenderTrust.

It is part of the government's drive to ensure that 100% of all government tenders are delivered electronically by 2002. Upon the successful completion of the pilot contract, OGC intends to roll out the piloted system across civil central government departments and agencies. The government expects the TenderTrust electronic tendering system to save the taxpayer up to £13 million over the next 4 years, and suppliers are expected to save some £37 million in the same period.

Until now the use of digital certificates and smartcards has been relatively limited because it has required a critical application to drive it. With e-tendering and the benefits it offers to all parties involved, its use is about to take off. Many more applications will follow, using digital certificates delivering trust in the transaction.

The quantity surveyor and e-commerce

For the purposes of this review, e-business has been broken down into the following:

* How the surveyor can participate in e-commerce
* Guidelines for the integration and adoption of e-commerce practice.

Quantity surveyors could never be accused of being luddites, and information technology already plays a large part in the development process. However, the DETR report *IT usage in the*

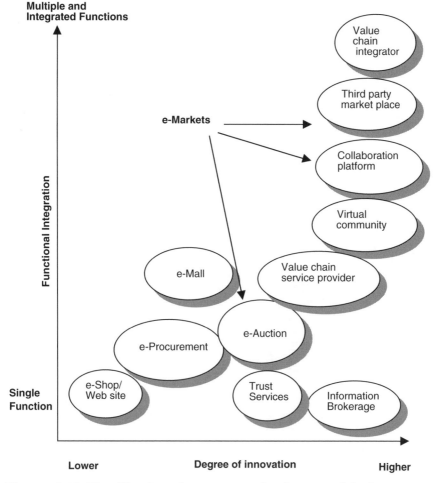

Figure 5.10 Classification of e-commerce business models (source: Paul Timmers, 2000)

Table 5.1 Levels of e-business

Level	1	2	3
Capability	Electronic mail & World Wide Web	Information management (electronic procurement)	Virtual organisation, e.g. Dell, Cisco, Amazon
What is needed?	Personal computer Internet access Software	Process redesign	Business redesign Culture change
Cost	Minimal investment Cost of running parallel systems	Change management	Change management
Benefits	Easy access to information Less information processing	Competitive advantage Streamlined business processes Alignment with supply chain partners	International competitiveness Best practice operations Robust relationships
Who benefits?	SMEs	Public sector Industry	Public sector Industry All members of the supply chain

Construction Team found that although the majority of information in the construction industry is created using IT, most is distributed in paper form. The report found that:

- 79% of specifications are produced electronically, but 91% are distributed on paper
- 73% of general communications across the design team are produced electronically, yet 85% are distributed on paper
- Only 5% of contractors' tenders are received in electronic format.

It is little wonder that it has been estimated that the European construction industry spends £300 million a year on couriers alone! In addition, BuildOnline claims that lost paperwork and lack of communication adds 20–30% to construction costs across the board. Tables 5.1–5.4 are included in order to give the surveyor some indication of the potential for the integration of e-commerce into day-to-day practice. Three levels have been proposed, based upon commitment (see Table 5.1), ranging from level 1 (see Table 5.2), requiring minimal investment but nevertheless still capable of producing tangible benefits and savings to both clients and business, through to level 3 (see Table 5.4), the virtual organisation, requiring total commitment and a high level of investment.

However, entering the world of e-commerce should be a considered business decision, and not a knee-jerk reaction to the sudden

Table 5.2 e-Business, level 1

Applications	Benefits
1. e-Mail	1. Real-time communication.
2. Marketing	2. Access to new markets on similar footing to larger organisations
3. Participation in electronic procurement/auctions	3. Clients may increasingly insist on electronic procurement
4. Access to databases, for example: • TED • Tenders Direct • Barbourexpert • Constructionline	4. Information on competitors/ market opportunities Sourcing using web pages

Table 5.3 e-Business level 2

Applications	Benefits
1. Fully electronic procurement including: • Pre-qualification • Comparing bids • Evaluation • Contract award • Archive information on bidders, including KPI rating • Online auctions	1. By far the biggest cost savings come from the reduced costs of creating and disseminating tenders; evaluation of bid responses, creating purchase orders and tracking progress Access to benchmarked records on contractors and supply chain
2. Fully electronic tendering including: • Tender submissions • Submission of bids • Online estimating • Exchange of information • Payment	2. Reduce wastage, increase profit margins Reduce errors
3. Project management • Dissemination of information	3. Effective supply chain management Real-time communication Integration of the supply chain
4. Enables virtual design teams • Collaboration with practices worldwide	4. Core personnel Contracted out work Part-time/ contract staff Contracted out staff could be located anywhere on the globe

availability of new technology. An existing bricks and mortar company's goals are different from those of an idea company. Probably already profitable, a bricks and mortar company will most often turn to the Internet to expand its markets, meet customers' needs and improve operating efficiencies in ways that are usually incidental to an existing business, rather than to

Table 5.4 e-Business level 3

Applications	Benefits
1. e-Construction packages including: • Project collaboration packages allowing construction professionals to access and amend project information	1. e-Construction packages: Alter and amend project documentation at minimal cost. Ensure that everyone works on the most up-to-date information. A repository for comments. Provide accurate audit trail. Compresses the construction programme. Few change orders.
2. Industry and supply side marketplaces	2. Real-time information of stock levels/delivery, etc.
3. Contractor consortia	3. Pooling and sharing of information.
4. Niche markets	4. Dispute resolution.

reinvent its business. The first step is to identify the business goals that can be served by using the Internet, and establishing these depends, in part, upon the nature of the business. For example, how large is the company? What sort of distribution chain does it utilise? Is its customer base static, or could it be expanded?

Even at the entry level, considerable market advantage can be gained through the use of e-mail and interactive web sites. The choice of partner can be crucial in maximising opportunities. For example, by working with existing operators, it is possible to start benefiting immediately.

A constant theme throughout this book has been client criticisms of the UK construction industry and its fragmented structure. Clients, professionals and the entire downstream supply chain try to cope with the challenge of operating in a highly fragmented industry, where the top five contractors own less than 10% of the marketplace. This fragmentation has over the years contributed to poor profitability and cashflow even for the major players, which has in turn prevented investment in new technologies. Add to this the constant demand by clients for added value, and the case for adopting e-commerce solutions and practices seems irresistible. As demonstrated in Figure 5.11, the biggest benefits of e-commerce are likely to come from the integrated supply chain, where information is freely available between clients, contractors and suppliers.

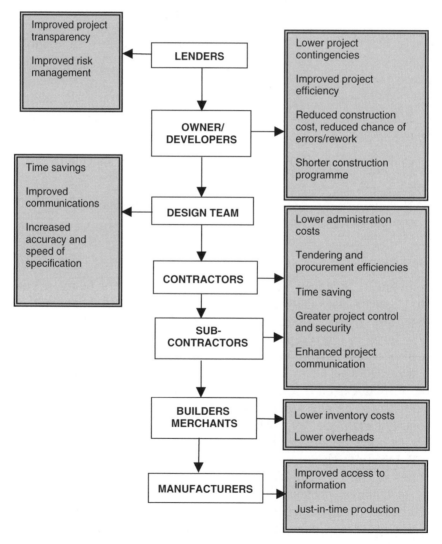

Figure 5.11 Construction supply chain benefits (Source: e-Business for the Construction Industry – BuildOnline, November 2000, and Kite).

This system illustrated in Figure 5.11, which is a level 2 application, allows savings and efficiency gains to be made through such measures as reducing the amount of abortive or repeat work carried out by the contractor and allowing manufacturers to employ just-in-time production techniques. Any changes that are made to the project information, for example alteration to the specification, are immediately communicated to all parties. e-Commerce allows true transparency across the supply chain, permitting the sharing of information between members in real time.

A further example is given in Figure 5.12, where the traditional lines of communication between the various supply chain members are illustrated. Without the presence of a hub to link

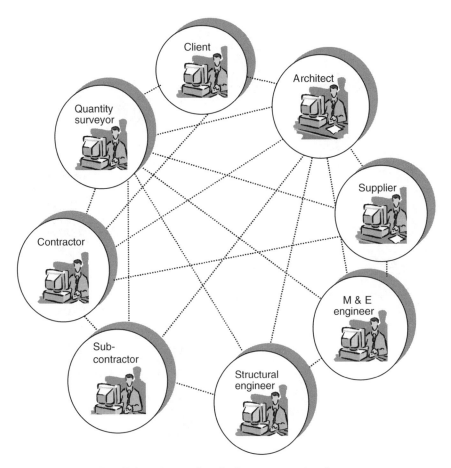

Figure 5.12 Traditional supply chain communication.

and co-ordinate the various parties, decisions can and frequently are taken in isolation, without regard to the knock on effects of cost, delays to the programme, impact on other suppliers, etc.

By contrast, the introduction of a collaborative hub or platform (e.g. the packages offered by the Building Information Warehouse, http//:www.biw.co.uk) permits decision making in the full light of knowledge about the possible implications of proposed changes. In addition it also permits specialist subcontractors or suppliers to contribute their expertise to the design and management process, thereby liberating the potential contained within the supply chain (see Figure 5.13).

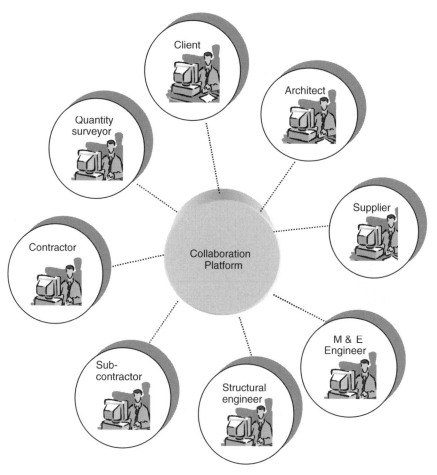

Figure 5.13 Collaboration platform communication.

Legal aspects of e-commerce

Beyond the basic business considerations, a host of legal issues face an industry or profession preparing to go online. This review of the implications of e-commerce for surveying practice cannot be concluded without mentioning some of the practical legal considerations that must be taken into account.

An aspect of e-trading that has caused considerable concern in both B2B and B2C sectors is the regulation of transactions in a global market, where organisations and governments can find themselves dangerously exposed in what has been described as a virtual 'wild west' environment. There is also the view, held by many in the IT sector, that the infrastructure is just not robust enough or sufficiently mature to do the job being asked of it as a result of the e-commerce hype.

The intersection of a global medium (like the Internet) with systems and legislations designed for the physical, territorial world poses many problems. Compared with other entities, the Internet has developed in a spontaneous and deregulated manner, and does not have a central point of authority. In the event of contested claims and possible litigation the fundamental problem of jurisdiction remains unresolved, as does the security of the systems, although (as previously described) steps are being taken to establish the integrity of cyberspace with the introduction of cryptography and digital signature services such as TenderTrust®. Its technical development has been guided by protocols established through bodies such as the Internet Engineering Task Force, but there has not been a central rule-making body that has exercised comprehensive legislative authority over the Internet, and there is unlikely to be one. The multi-jurisdictional and multifunctional nature of the Internet means that inevitably many different interests in many different parts of the world will be concerned with any endeavour to formulate specific policies. Even in the European Union, the suggestion contained in the Commission's draft directive that e-commerce should be governed by the law of the country where the service provider is established has been questioned by consumer groups that want the local law where the website is accessed to be given priority.

Another crucial area of concern, primarily for governments, is e-tax, or the tax treatment of online transactions. At present the volume of e-transactions means that the fiscal implications are modest. However, if predictions of growth are to be believed, the questions of which government is to tax such revenues and how could be very important. Within the OECD area, views diverge; in the USA the belief is that e-transactions should not be taxed, while in the European Union the view is that VAT should be levied on e-trading. The law in the field of e-commerce is continuously and rapidly developing, as numerous drafts pass into the statute books. The evolution of technology also means that legislation must be updated and requires constant review. Organisations need to be aware of both the current and the prospective impacts of legal provisions.

The principal regulatory concerns are focused on four areas; online contracting and security (both dealt with previously in the review of TenderTrust's encryption and private key services), regulation/jurisdiction, and intellectual property protection.

The implementation of electronic keys is dealt with in an EC Framework Directive to be implemented by July 2001. When originally drafted, the UK e-Communications Act 2000 contained powers to be vested in law enforcement agencies, requiring disclosure of electronic keys to de-crypt information where necessary. This provision caused considerable debate, as it was seen by many as a major barrier to the promotion of the UK as a favourable environment for e-commerce. It was suggested that the Home Office had hijacked the bill, and eventually the government was forced to delete the provisions, although in practice they were only moved to the Regulation of Investigatory Powers Bill, which is now on the statute books. The UK e-Communications Act is therefore now quite simple in that it allows for:

- The introduction of a new approvals regime for providers of cryptography services
- Electronic signatures to be admissible in court
- The updating of existing legislation to allow the use of electronic communications.

In the early days of e-commerce it is true to say that the legislative framework was lagging behind business practice. How can an organisation be sure that information being transmitted electronically is secure from its competitors? The major legislative instruments in e-commerce law are:

- The UK Electronic Communications Act 2000
- The Electronic Commerce Directive (00/31/EC) – implementation planned by January 2002
- The UNCITRAL Rules – still at the negotiation stage.

However, this legislation only applies to services supplied by service providers within the EU. Countries outside the EU are covered by UNCITRAL rules. The e-Commerce Directive establishes rules in the fields of: definition of where operators are established; transparency obligations for operators; transparency requirements for commercial communications; conclusion and validity of electronic contracts; liability of Internet intermediaries; and online dispute settlement. Put simply, the Directive states that the service providers are subject to the law of the member state in which they are established, or where the ISP has its 'centre of activities'. B2B contracts are of particular importance, as national laws govern the main aspects of contract law, and what constitutes an offer or an acceptance varies from country to country – for example, at what point is the offeree's acceptance communicated to the offeror? Communication by web site is instantaneous, whilst e-mail is not. The possible scenarios include:

- The offeror failing to collect e-mail from the server
- Failure of the ISP.

The European Directive 00/31/EC attempts to clarify the situation by stating that the contract is concluded when the offeree is able to access the offeror's receipt of delivery. Unfortunately this clause does not cover the position of, say, an invitation to tender, which is an invitation to treat.

In general, the parties to the contract should agree by what is called private autonomy as to which country's law of contract is to apply.

Conclusion

There can be no doubt that the future for both business in general and for the quantity surveyor in particular is as part of the digital economy. In the late 1990s revolutionary new business models were set to destroy old economic values, but only 2 years later the talk of the collapse of the new economy was just as overstated.

However, to equate the downturn in the e-economy with the demise of the Internet is, as with the pundits who proclaim the death of the quantity surveyor, a gross exaggeration! The market is maturing and, as has been demonstrated in this chapter, the march towards the digital economy is unstoppable. Of late the headlines have been made by organisations that are bucking the trend and still continue to invest in e-commerce. There are real benefits that can be brought to a fragmented industry by e-commerce – a survey by King Strurge and H+H Marketing in mid-2001 showed that 73% of property investment companies either are planning or have an e-commerce strategy. The demand now from clients is for increased bandwidth and fibre-optic technologies to allow greater speed of connection to be available in their buildings. The old maxim that 'Location, Location, Location' was the critical success factor for a development's success has been replaced by 'Bandwidth, Bandwidth, Bandwidth'.

Despite all the pitfalls, both technological and legal, outlined in this chapter, what are the critical success factors to be considered by an organisation still determined to be part of the digital revolution? Assuming that the starting point for a new e-commerce venture is its ultimate success, the points to consider are as follows – although these will vary in line with the level of entry (see Table 5.1).

- *Content*. This should be a unique and/or innovative product or service that exploits the electronic environment and delivers added value to potential users. In addition, the platform must be capable of attracting sufficient clients to generate the cash flow to repay the start-up and running costs. An example is dispute resolution online.
- *Community*. There must be the ability to build up a critical mass of customers/business partners for the venture, which will translate into sales/cost savings to cover the initial investment. It is the inability to establish a community of clients that has caused so many dot.coms to fail.
- *Commitment*. Clear objectives are necessary. These are most clearly demonstrated by a defined business case for the e-commerce venture, but at the very least there must be a clear idea of the objectives and a demonstration of strong motivation for using the Internet. An example is just-in-time production for building materials – a large volume of curtain walling could be produced only when it is required for incorporation

into a project, and the manufacturer in return can tap into a global supply base for raw materials.

- *Control.* This is the extent to which e-commerce is integrated with the internal business process, enabling the organisation to control all aspects of its business and handle growth and innovation.

The next 5 years

A report published by the Building Centre Trust in December 2001 and entitled *Effective Integration of IT in Construction* concluded that information technology held a key role in the implementation of integration, collaboration, teamwork and partnership in the construction industry. The report included a model for the future integration of IT into the procurement of construction projects, as illustrated in Figure 5.14.

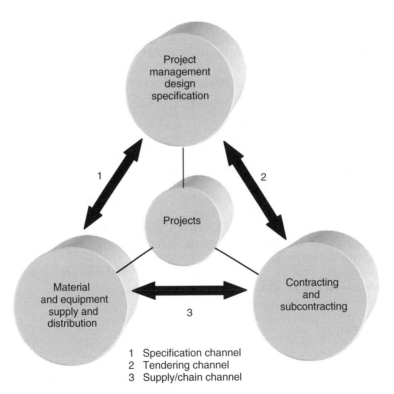

1 Specification channel
2 Tendering channel
3 Supply/chain channel

Figure 5.14 e-Business development model (source: Building Centre Trust).

The report identified three areas where rapid development and improvement are predicted to take place over the next few years, all of which are mainstream services for the quantity surveyor; tendering, the specification of materials and products, and supply chain integration.

References

Building Centre Trust (2001). *Effective Integration of IT in Construction.* Building Centre Trust.

Hobley, C. (2001) *Just Numbers. Numbers on Internet Use, Electronic Commerce, IT and Related Figures for the European Community.* Published by The European Commission's Electronic Commerce Team (INFSOC DG).

Timmers, P. (2000) *Electronic Commerce – Strategies and Models for Business to Business Trading.* Wiley & Sons Ltd.

Further reading

Betts, M. (1998). Information technology is fundamental to everything Egan is talking about. *Building,* 16 Oct, p. 73.

Building (2000). Industry e-commerce to hit £67bn by 2004. Building, 6 Oct,

Chappell, C. and Feindt, S. (1999). *Analysis of e-Commerce Practice in SMEs.* Kite.

Davidson, S. J. and Kapsner, M. (2001). Bricks and mortar to the Internet. *Journal of Internet Law,* Mar,

Department of Trade and Industry (1999). *Building Confidence in Electronic Commerce – A Consultation Document.* HMSO.

Government of New South Wales (1999). *Electronic Procurement – Taking up the Challenge.*

Khatri, M. (2000). The construction industry hits the internet. *Chartered Surveyor Monthly,* May,

Schulze, C. and Baumgartner, J. (2000). *Don't Panic ! Do e-Commerce – A Beginner's Guide to European Law Affecting e-Commerce.* The European Commission.

Thompson, B. (2001). Numbers that will unsettle the e-sceptics. *Estates Gazette,* 21 Jul,

Web sites

www.buildingcentretrust.org

www.dti.gov.uk

www.ecommerce.gov

www.e-envoy.gov.uk

www.europa.eu.int

www.ogc.gov.uk/

www.ogctendertrust.com

www4.gartner.com

www.buildonline.com

6

New technology – opportunity or threat?

Graham Castle MSc FRICS

Introduction

During the preceding chapters the opportunities for quantity surveyors to develop new expertise in areas that are either available now or emerging have been described. Many of these opportunities depend on the adoption of new technologies. This chapter reviews other industries' experiences of integrating new technology into existing work practices and processes, and concludes with a strategy for successfully integrating new technology into a surveying organisation.

A bottom-up approach

Technology has had a profound impact upon other industries over the last decade, but not on the construction industry; why is this so? This chapter sets out to answer this and other important questions, such as how to employ new technology in the workplace, and where to use it.

The adoption of new technology has changed the banking industry beyond all recognition and has led not only to the development of new services but also, more importantly, to new ways of working. Virtual money may not yet be a reality, but the virtual bank is. Bank managers have become an endangered species, and you can make payments and withdrawals 24 hours a day, 365 days a year using an automatic cash machine or the Internet. Indeed, it is possible to do all of your banking today without having any face-to-face contact with another human being.

However, the banking industry is not an isolated example. Twenty years ago, the newspaper industry was, if anything, even more conventional in its ways than the construction industry is today. It was only after a long, protracted and violent industrial dispute that employers were able to introduce new working practices based upon new technologies that enabled digital publishing. This exemplifies the fact that new working practices and the allied new technologies are not always welcomed into the workplace by everyone. In the newspaper industry, the emphasis was on the introduction of new digital technology and the elimination of historical working practices, combined with the irrelevance of some traditional trades. Another example of the adoption of new technology is the automobile industry, discussed in Chapter 1, where once again the introduction of new working practices such as computer-aided design and manufacture, just-in-time materials deliveries, electronic ordering and bill payment has enabled manufacture on demand. Here the emphasis has been upon the integration of design and manufacture, supply chain management, and the introduction of robotics.

These examples clearly show that in order to gain the full benefits of new technology, working practices have to change. Therefore, it seems likely that the quantity surveyor is going to have to do more than merely become competent with a spreadsheet. It is also worth noting that all of these so-called revolutions in other industries occurred before the age of the all pervasive Internet!

Historically, quantity surveyors have not been slow to adopt technology. They were amongst the forerunners in the construction industry to adopt personal computers and software such as spreadsheets, bills of quantities production software and word-processing. However, one problem associated with the introduction of computers into the workplace was that technology competence tended to decrease in direct proportion to seniority within an organisation. This lack of knowledge of new technologies on the part of senior surveyors, the very people who are identifying the business objectives and managing the firm, led to scenarios where investments in new technology were driven from the bottom up within companies. Typically, a graduate surveyor would identify a task that could be made more efficient or less tedious by the application of a new technology. This led, often after considerable persuasion, to a purchase being made to satisfy the individual user's need. Another typical scenario was that a nearby competitor would acquire some new technology and it was then felt that the firm should do likewise so as not to be left behind. This approach led to islands of computing being established within firms, where new technology was acquired upon a task basis with little or no attention being paid to the overall strategy of integrating new technology into the workplace for the benefit of the business overall. Not surprisingly, many of these investments failed to provide the expected returns, and in some instances they were even counterproductive, leading to a fall in productivity.

Although the examples given at the start of this chapter regarding acquiring new technology are all on a grander scale than that of the typical quantity surveying firm, they are all typified by being championed at the highest levels of management within their respective organisations. This does not mean that senior managers in the banking, newspaper or the automobile industries are more computer literate than senior surveying managers. It does, however, identify that they recognised how new technology could be allied to their business objectives, and then championed that cause through thick and thin, sometimes ruthlessly, to achieve their aims. The bank clerk, the journalist or the assembly line worker did not propose to senior managers how their daily work could be made more efficient by the adoption of new technology! Later in this chapter we will look at how surveying firms should go about identifying, acquiring and employing new technology in the workplace.

Is construction unique?

Is it reasonable to compare the property and construction industries with banking, newspapers and automotive industries?

Banking, just like construction and property, is a service industry; however, its product is less tangible than a building and is also much more standardised, in that the same products will be offered as a service nationally throughout the bank's marketplace. Banks are also very profitable, and are large commercial organisations employing thousands of staff. In addition, they have a large base of customers and principally deal in financial data. It was the computer power of storing, sorting and manipulating data that led to the original introduction of computers into the banking industry. Therefore, because the industry had large profits to reinvest in their business activities, a standardised product, hundreds of thousands of data sets and a very structured data format, it was very easy for new technology to be introduced.

The newspaper industry introduced new technology for a different reason – introducing new working practices. Here the problem was archaic working practices that had changed little from the days of William Caxton, together with very powerful trade unions that were opposed to the introduction of the new technology. The trade unions were acting in what they thought was their members' best interests in trying to protect the livelihoods of their members. However, in refusing to acknowledge or even consider any new working practices, they eventually succeeded in making the skills of their members irrelevant to the modern newspaper industry. New technology was introduced by being championed by the newspaper owners, and this led not only to new working practices but also to new products. The lesson here is that the advance of technology cannot be ignored or reasonably repulsed, and it is inevitable that some traditional skills will become redundant in the process. Furthermore, to be successful, the introduction of new technology requires a champion at senior management level. Once again the newspaper industry is an example of an industry comprising very large organisations with high levels of profitability and a large workforce.

The introduction of new technology into the automobile industry was for different reasons again. First, the aim was to integrate design and manufacturing via CAD/CAM and robotics; secondly, it was to allow new working practices associated with supply chain management. Here, once again, the introduction of new technology led to the development of new ways of working. Senior management again championed the introduction of new technology, and the firms in question were profitable, had huge annual turnovers, employed thousands of staff, and manufactured very similar products in vast quantities under factory conditions.

These examples from other industries demonstrate that, to introduce new technology into the workplace successfully, it is necessary to:

- Champion new technology at a very high level in the company.
- Ensure that the new technology is focused upon helping the firm achieve its business objectives.

- Re-engineer existing business processes, often leading to the introduction of new ways of working, new services, and the redundancy of some traditional skills.

The construction and property industries, on the other hand, have quite different characteristics (see Table 6.1) in that they are typified by:

- A proliferation of small firms forming temporary alliances upon a project basis for the duration of that project.
- The separation of the design and construction activity.
- Uncertainty due to variable demand, onsite construction hampered by inclement weather and the lack of a standard product.
- Fierce competition for work that creates low levels of commercial profitability.

Table 6.1 Characteristics of construction and property industries

Characteristic	*Example*
Fragmentation	Design and construction processes separated. Any one project has many stakeholders. Industry structured upon basis of project teams that are temporary alliances.
Uncertainty	Inclement weather can adversely affect progress on site. Very competitive environment. Process accommodates design changes during the construction phase. Every building is unique to a lesser or greater extent. Buildings are becoming increasingly more complex. Nomadic labour force.
Poor communications	Missing or conflicting design information at the construction stage. Late transfer of information between design team and the contractor, and contractor and subcontractors. Conflicting information sources, e.g. between bills of quantities and project specification.

- Each product being, to a greater or lesser extent, unique.
- Hundreds of stakeholders in any one project, many of whom may have conflicting business objectives.
- Displaying considerable resistance to any change in working practices.

Comparisons between the construction and property industries and those of banking, newspaper publishing and the automotive industry are therefore not entirely reasonable, but should not be completely dismissed. The principal differences are that the property and construction industries lack a champion to lead the introduction of new technology, and that the average construction and property firm is very small – typically employing less than 12 staff – as well as unprofitable, many achieving less than 3% profit margins. A further complication is the organisation of the construction industry on a predominantly project basis, with a proliferation of stakeholders, which leads to poor communications, misunderstandings and errors. Where then might a champion be found? Given the poor success rate of acceptance of technology into the construction and property industries, you would be forgiven for thinking that none exist.

The most obvious champion is the senior management of any large organisation active in the construction and property industries, such as a large contracting organisation or multi-professional practice, and there are indeed examples of this. The new technology in these instances is usually associated with company-based information systems introduced to achieve company objectives. An example of this type of system is the Franklin + Andrews CCMS System, which is discussed later in this chapter. These successes are due to the fact that they are contained within the boundaries of one company or organisation, where common standards can be introduced and enforced, and that large companies are more likely to have both the vision and the funding to support the appropriate introduction of new technology. Typically, however, these instances do not display the characteristic of introducing new working methods, but are targeted upon the achievement of a company's business objectives. The other obvious champions are the larger clients – those who are constantly commissioning building works. Most prominent of these is central government. The government has commissioned various reports, e.g. *Rethinking Construction*, into the operation of the construction and property industries, most of which have recommended greater use of new technology allied to the introduction of new working practices. Interestingly, unlike the previous example these have been targeted at the project level and are geared to achieving the client's objectives. The Office of Government Commerce's initiative to introduce the widespread use of web-based systems for procurement, discussed in Chapter 5, is an example of this approach.

New technology and information

Just because something is technically possible it does not mean that it should be adopted in the workplace, and there are many examples of new technology being employed for its own sake. Such instances are doomed to failure, as they are not allied to supporting the achievement of the company's business objectives. New technology is often referred to as information technology; note that the term 'information' precedes that of 'technology'. This is indeed the correct relationship, as new technology or information technology should always support the company's information systems. It is the information systems that are important, more so than the supporting technology. The purpose of any information system is to ensure that appropriate accurate information is made available to the correct person timeously in order that a decision can be made. Information itself is merely processed data, and data are facts that in themselves have no apparent meaning. The problems associated with the building of information systems revolve around who needs what information when, how to acquire appropriate data, and how those data can be processed into the requisite information.

Information systems are the product of systems analysis, and it is the systems analysis that leads to the business re-engineering and new ways of working described earlier in this chapter. Systems analysis concerns itself with the optimum way of achieving a task, and not the recording of how a current task may be being performed. It is the missing information systems that have led to the many failures in investment in new technology in the construction and property industries. This is also partly responsible for the failure to adopt new working practices, as most investments in new technology based information systems have been based upon the automation of existing working practices.

This has resulted from the bottom-up approach taken by many investments in new technology in the construction and property industries, where investments have been made to support individual information needs rather than those of the company overall. It is also further compounded by the identification that, in reality, two mutually supportive information systems are required to assist build environment business, namely:

1. A company based information system to support the company's objectives
2. A project based information system to support the client's objectives.

This in turn leads into the very important concept of the need for integration of information systems in the construction and property industries. Earlier in this chapter fragmentation of the industry, which is typified by the large number of stakeholders involved in any one project as well as the preponderance of islands of computing or islands of information, was identified as a problem to be overcome. The obvious solution to these problems is integration. Unfortunately, the adoption of new technology does not automatically lead to this integration of systems.

Integration and new technology

There are numerous and different views of what the integration, in terms of construction and property information systems, really means. The simplest view is that it merely refers to the integration of one new technology-based information system with another to eliminate the islands of information. A more radical view is that it relates to the integration of numerous new technology-based information systems in order to achieve the goal of developing project based information systems, through which project stakeholders can share information and work in a collaborative way – sometimes referred to as computer-integrated construction (CIC).

New technology is something of a paradox when considering how it can be employed to achieve integration of information systems in the construction and property industries. The simplest route is where everybody uses the same computer operating systems and software. However, even this route can lead to problems if people are using different versions of the same software, have set up the software to operate with differing default settings or, in the case of CAD, simply adopted contradictory layer arrangements. This can of course be simply remedied by enforcing a framework or structure regarding how the new technology should be employed. This requires a champion to drive the concept forward and ensure that the framework is adhered to. Examples of this approach to integration based upon new technology are most often found within one firm or a single organisation, and consequently this approach is well suited to the company-based information systems identified earlier in this chapter.

A more common instance is where integration is required across a range of operating systems and a variety of disparate software. The solution here lies in the adoption of either neutral file formats or open architecture computing. Neutral file formats have long been the Holy Grail sought by organisations and governments as the tool that will enable easy transfer of information between differing computer systems, especially CAD systems. Early attempts at solving this problem were based upon translators that were used to convert data from one system to another and *vice versa*. This was not really a feasible solution as the number of translator applications required grew exponentially to the number of CAD systems in use, and it was obvious that another solution was required. In Europe this took the form of the Initial Graphics Exchange Specification (IGES) and, later, the Standards for the Exchange of Product Data (STEP). STEP, now an international initiative, is concerned with the exchange of product data between computer-based information systems. In the USA, a similar initiative took the form of the Product Data Exchange Specification (PDES), which in turn led to the development of Product Data Modelling (PDM). PDM is yet another international initiative concerned with the use of new technology for the representation and exchange of product data. It is based upon four other international standards: Electronic Data Interchange (EDI; discussed in Chapter 5); ISO 8879 – Standard Generalised Markup Language; ISO 10303 – STEP; and ISO 13584 – Parts Libraries.

The most recent initiative is that of Computer Acquisition and Lifetime Support (CALS). This has its origins in the US defence industry and is now also a UK government-sponsored initiative, and includes the construction and property industries within its scope. CALS is also a combination of existing and emerging standards, including EDI and STEP. All of these initiatives are government sponsored, are targeted at achieving integration at the data level, and are not specifically targeted at the construction and property industries but at all industries – especially the manufacturing industry.

Given the predominance of small to medium-sized enterprises (SMEs) in the construction and property industries, and the complexity of the solution in relation to the type of project that constitutes the day-to-day workload of these firms, it is obvious that the generic solution to data integration does not lie down this avenue. A more likely solution, and one that is commonly adopted by SMEs, is that of adopting the proprietary neutral file formats that are available commercially in the new technology domain. These provide only a partial solution to the problem of integration, but are generally free, commonly used, and reasonably well understood by users. However, they cannot really be compared to the former solutions in scope, intent or usefulness. Examples include the use of text file and HTML files for text-based documents, comma delimited files or space delimited files for numeric data, and DXF, the AutoCAD neutral file format, for CAD files. Although readily available, they provide an incomplete and imperfect solution.

An alternative solution that is showing promise is that of Industry Foundation Classes (IFCs), which are being developed by the International Alliance for Interoperability (IAI). This is an international non-profit making association of companies and researchers active in architecture and engineering construction. The IFC development workload has been devolved to the members on a national basis, each nation being set an objective that contributes to the overall international goal. IFCs that work at the project data level are associated with object-oriented programming, which is an entirely new and radical approach to the development of computer programmes and CAD systems in particular. In simplistic terms, the basic object-oriented concept is that the world is made up of objects, that these objects can inter-react with each other, and that they can be simulated in the software programming. Consequently, object-oriented CAD software would be able to create dynamic models, or models that can react to their environment or usage. Integration is at the core of the IAI initiative, and IFCs are merely an attempt to define objects in the construction and property industries, which can then be employed by software designers to develop software for the industry that is capable of displaying 'interoperability' or integration. Software built to comply with IAI IFCs probably stands a much greater chance of success and adoption across the industry than STEP or CALS.

The last possible solution to the problem of integration is based upon 'open systems computing'. This relates to the desire for hardware platform and operating system independence, and is commonly interpreted today as meaning Internet technologies. Many of these tools are based upon the standard generalised mark-up language (SGML), which is in effect a set of rules for constructing other Internet languages. There are many of these and they are constantly under development, so cannot yet be considered mature. The most common example is hypertext mark-up language (HTML). HTML, although useful as a publishing medium, has limitations for business use, and amongst its faults are the fact that the designer has little control over what the finished page looks like on the computer screen, and also that it is really only useful for text-based information. A more valuable example for the built environment is virtual reality mark-up language (VRML), which is an Internet standard for 3D graphical information. This enables the viewing of 3D graphical models over the Internet.

Extensible mark-up language (XML) is a simplified form of SGML, and is a set of rules for creating mark-up languages. XML-based languages, however, are not concerned with appearance, but rather contain information about the logical structure of the document. The important thing about XML-based languages is that they more readily enable the transfer of documents between computer systems without those documents losing their structure. XML could be used, for example, to send an invoice between one computer and another, despite both computers having different hardware architectures and operating systems. Indeed, XML shows every promise of more readily and easily enabling electronic data interchange (EDI) to be achieved than do current technologies.

The solution to the built environment problem of integration may well lie within a combination of these Internet technologies.

New technology and the knowledge worker

Few people in the construction and property industry actually earn their living by physically creating a solid product, such as the building itself or any of its many components. Most jobs instead involve the management or processing of data, and people working in these fields are commonly called information workers. Information workers can be further sub-classified as data workers, as people who store, retrieve and manage data, or as knowledge workers – those who create new information or knowledge. Professionals within the built environment are knowledge workers because they add value to the product via the application of their knowledge and expertise in order to create new knowledge about the project – e.g. the architect produces the conceptual design, and the quantity surveyor the cost and procurement advice. New technology is capable of supporting both the data worker and the knowledge worker.

The object of providing new technology for knowledge workers is that by doing so their productivity will be improved. This is of prime importance to most professional firms, as the knowledge workers are also the fee earners. Any investment in new technology here should aim to integrate the knowledge and expertise of the knowledge worker into the business, rather than create a desktop tool for the sole use of the individual fee earner. Once again integration with other office systems is all important, and this is normally interpreted as being the easy transfer of information between knowledge work systems or data stores. This invariably necessitates the use of networks and/or intranets, and perhaps also the Internet. Examples of knowledge work systems in the built environment include computer-aided design packages (CAD) for architects, and estimating and bills of quantities production systems for quantity surveyors. Other examples include modelling and rendering software and analysis software (e.g. structural or environmental analysis software). It is worth noting that these tools in themselves are incapable of doing anything very much, and it is only through the application of the workers' skills, expertise and analysis of the results produced that new knowledge or information is created.

Recent trends in built environment knowledge work systems display this desire for the systems to be integrated into the business and/or the project environment rather than being seen as a desktop tool for the sole use of one person. ArchiCAD V7.0

is marketed as being a network-based project design tool that can be used by a team of designers to develop and manage a project design and documentation. This tool goes way beyond the normal concept of CAD as being an electronic drafting board. For example, Q-Script, a tool for the quantity surveyor, displays similar characteristics along the lines of enabling collaborative project working upon a team basis and, being a networkable product, also enables the management and documentation of the surveying services on a project.

Other new technology tools, although not directly helping the productivity of the knowledge worker, can be employed to improve collaboration and communication between workers. These tools are concerned with enabling communication and collaboration generally, and are not environment specific. Examples of these include electronic group diary and schedule systems, contact management systems, document management systems, company intranets and e-mail systems, publishing systems, and even simple bespoke databases and word-processed documents. All of these can be used to share and distribute knowledge. However, as before, investment in these must be appropriately focused upon the company's business objectives in order for any real benefit to be realised. The purchase of any of these systems does not in itself guarantee increased productivity and efficiency. e-Mail systems and the Internet in particular can be considerable time wasters if not properly integrated and controlled in the workplace.

New technology and graphical information

Graphics play a very important role in both communication and the transfer of information in the property and construction industries. These graphics can take many forms, such as freehand sketches, detailed technical drawings, and as built drawings. Traditionally this graphical information has been passed between project stakeholders in the form of 2D drawings. Techniques have even been developed to represent 3D views of buildings on a 2D page (e.g. isometric and oblique representation). Even relatively small projects can result in the production of tens or even hundreds of drawings.

The project drawings are merely a proven method of recording and communicating project information. The conventions for the production of project drawings have developed over many years but, despite having been in use for many generations, their production is not problem free, even today. The production of project drawings is a collaborative task, with many different individuals and firms contributing to the final product. Almost without exception, these drawings are produced today using a computer-aided design (CAD) system.

The backbone of all CAD software is a database, although most users are not aware of this fact. The structure of a CAD database is different from that of a normal relational database such as MS Access, which is made up of fields, records and tables. A CAD database is just a well organised list of data, and is capable of storing two kinds of information; the geometry information required to produce the CAD drawing or model, and non-graphical information inserted by the user. The technique for inserting non-graphical information into a CAD database is to add the information to blocks via the block attributes facility (Figure 6.1). Historically designers have not commonly adopted this facility, since it is perceived as being non fee earning work and for the benefit and use of other professions, and has therefore largely been ignored. The advent of object-oriented CAD software will to some extent solve this problem, in that much of the data in an object-oriented CAD database will be produced automatically without the need for the CAD user to insert information manually.

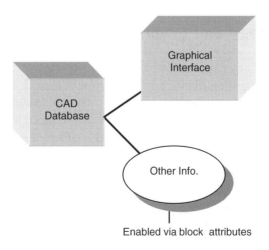

Figure 6.1 CAD structure.

However, if CAD is to form the basis of a project information data model capable of supporting collaborative working, the capability of the CAD database to store and retrieve non-graphical information must be used to its full potential. How, then, might block attributes be used to store project data in the CAD database? The most obvious application is the recording of component attributes (e.g. information associated with doors, windows, trusses etc.). Block attributes can also be used to store pre-design information associated with the project brief, such as information concerning an employee workstation or office in an office development. Indeed the opportunities are only limited by imagination and ability, and can include the incorporation of CAD data into reports, specifications, schedules and other construction related documents (Figure 6.2).

Block attributes can usually be displayed on the CAD drawing that is being worked on, or extracted into a data file that can be subsequently imported into another software application, such as a spreadsheet or a word processor. More importantly, block attributes can also be linked to a field in external relational databases via the adoption of Structured Query Language (SQL). An example of this technique would be to link a door block in a CAD drawing to an external database that contained other fields relating to door dimensions, types,

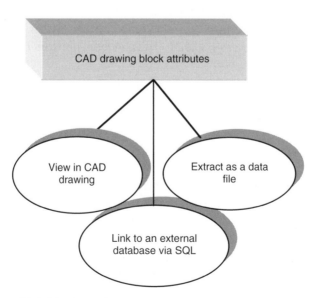

Figure 6.2 CAD block options.

hardware etc., thereby enabling the production of the project door schedule. This leads to the concept of the CAD drawing becoming an intelligent interactive drawing that does not require the CAD user manually to insert all the project data associated with block attributes. It also exemplifies the basic computing rule of only ever having to insert data into a computer system once.

CAD is therefore capable of supporting project information databases. All projects have a need for information management, and any supporting information system must be focused on the following project based critical success factors (CSFs), all of which are interdependent:

- Timely completion of the project.
- Completion of the project within budget.
- The quality of the finished product fully meeting the functional requirements of the client.

A core objective of any information system is to ensure that the correct information reaches the appropriate person when it is needed. In the construction industry, communication lines are often regulated by the form of contract adopted rather than by the optimum information system architecture. Any built environment project based information system must be capable of supporting all three CSFs of time, cost and quality. Traditional built environment project based information systems only support one of the core CSFs; that of completion of the project within budget.

Project information can be categorised as follows:

- Commercial
- Technical
- Managerial.

Each category satisfies a different project need and has unique characteristics. A project-based information system would need to accommodate all three formats of information and recognise that administrative functions are interwoven with commercial, technical and managerial information needs.

Commercial project information (e.g. accounts, payroll, tax returns, invoices, labour returns, plant returns, delivery notes etc.):

- Needs to be 100% accurate
- Is not time-critical (in terms of project CSFs)
- Must be 100% auditable
- Is historical in nature
- Can be used as data to predict future events (project control)
- Is largely administrative.

A CAD-based project information model has little to offer in terms of support for commercial-based project information, which in many respects reflects more the requirements of a company-based information system. The information technology tools best able to support this model lie outside the functionality provided by a CAD system even when linked to other systems such as a relational database.

Technical project information (e.g. client briefings, design guides, project drawings, project specification, bills of quantities, engineering calculations, environmental impact analysis, contract conditions, variation orders etc.):

- Needs near 100% accuracy
- Is approaching 100% time-critical
- Should be auditable
- Is procedural driven
- Is directly related to project control mechanisms (CSFs)
- Results in a project 'knowledge base' (expertise and experience).

This category is the domain in which a CAD-based project information system has most to offer, even without the advent of OOPs-based CAD software. The techniques involved were discussed earlier in this chapter, and their application is limited only by imagination and technical competence. Research in this area is being targeted at the development of intelligent CAD software that guides the user through technical or professional processes and integrates either technical or professional knowledge with the CAD software. An example of this would be a CAD system that advises on fire escape regulations as the design develops, or the building regulations in respect of a stair design.

Managerial project information (e.g. project time, cost, quality, feasibility factors):

- Requires a degree of accuracy
- Is concerned with trend analysis

- Has no audit requirement
- Has a significant project control requirement
- Is 100% time-critical
- Is forward looking
- Is management based.

Much research is currently being devoted to the development of CAD-based information systems that are capable of supporting managerial project information, an example being the linking of CAD models to project-based planning systems. Most of these initiatives fall under the umbrella of Computer-Integrated Construction (CIC). There are also a number of examples of this use of new technology being introduced into the workplace; however, these are normally in instances where design and construction fall under the umbrella of one company, e.g. a design and build contractor, or a multi-professional firm.

CAD-based project information systems are therefore most able to support the project requirements for technical information systems, and show some promise in being able to interface with managerial information systems, especially in circumstances where there is less project fragmentation. Yet again they are based upon integration of information and systems, and to achieve fully the benefits on offer will require the development of new working practices, as was the case in the newspaper industry.

New technology and commercial advantage

Can the adoption of new technology assure commercial advantage?

The adoption of new technology is often thought to be associated with commercial advantage, and there is a huge volume of published material on the subject. One thing is sure; what is done today (should it prove successful), the competition will copy tomorrow. Consequently, commercial advantage associated with new technology is often short lived. Companies in the built environment are also difficult to differentiate from each other; they all provide a similar range of services, the setting up of new firms is not prohibitively expensive or difficult, and none have the authority actively to influence their own supply chains. These characteristics, along with the high level of fragmentation in the industry, make it very difficult to adopt new technology with the aim of achieving any lasting commercial advantage. Furthermore, a large number of SMEs active in the built environment do not have the time, resources or vision to investigate or experiment with how technology could be adopted to gain any commercial advantage. A more common scenario is that technology has to be adopted to enable them to compete with rival firms that have already attained some advantage. Motivators of this type are unlikely to result in either the successful introduction of new technology into the workplace or the attainment of the expected benefits.

Commonly, new technology is adopted to achieve efficiency gains and thereby a cost advantage over competitors. Other businesses, fewer in number, have adopted technology to develop new services that enable them to differentiate their services from rivals. A few have adopted technology to enable the adoption of new working practices or structures, and it is perhaps these that will prove to be the most enduring and valuable. The successes achieved by the banking, automotive and newspaper industries all resulted from the adoption of new technology allied with new working practices and organisational structures. The adoption of new technology itself is therefore unlikely to result in any lasting commercial advantage, although failure to adopt technology commonly used by competitors could lead to a failed business. The most likely way in which technology could be adopted to achieve lasting commercial advantage would be where it is adopted to support new services, enable the restructuring of the organisation, and/or introduce new working practices.

New technology and people

For new technology to be welcomed into the workplace, the users must accept it. To be accepted by the users, technology must be seen as benefiting them by making their jobs less tedious or giving them new skills and responsibilities. It is also wise for the potential users of a system to be involved in its selection, development and introduction, as this engenders a feeling of ownership. Any attempt to impose a system upon users unilaterally is unlikely to prove to be a rewarding experience.

The introduction of any new system into the workplace will require staff training. Training is vitally important, and many new technology systems fail to produce the expected benefits owing to a lack of formal training. It should not be seen as an afterthought but as an essential ingredient, and it is quite likely that training will be required over an extended period of time, rather than just a few days. It is also good practice to have staff train each other, and for the user knowledge to cascade down through an organisation. There should always be at least two people with intimate knowledge of any one system to ensure continuity of expertise in the event of illness or staff changes. Initially training must take place away from the day-to-day workplace pressures, although latterly it can be integrated into the daily routine.

The introduction of any new technology system will inevitably result in a fall in output as staff become familiar with the new equipment, and it is likely to be some months before performance recovers to its former levels, and perhaps a year or more before any productivity benefits become apparent. Staff should work their way through the performance levels, from becoming familiar with the technology to becoming competent users and ultimately to becoming innovative with the technology. Many firms fail to progress beyond the stage of becoming competent with the technology. It is a good idea for each member of staff to become the guru with regard to any one application and then to act as a focal point for queries relating to that application.

Training is not free, and it is not unusual for the cost of training, including the associated fall in staff productivity, to exceed the investment made in purchasing the new technology itself. Training should also form part of each system upgrade, and not be provided only when the system is first introduced into the workplace.

How best to employ new technology

The stimuli for employing IT in the workplace are not always positive in nature, and are often associated with what clients expect of your firm, the need to collaborate with others in a specific project team, staff expectations, or simply the need to remain competitive. As identified earlier, investments made upon these bases are unlikely to lead to success. It is not the IT itself that is important, but how it is employed within your organisation that is vital.

In giving advice upon how IT should best be employed in an organisation, consideration has to be given to these factors:

* The size of the firm (SME or other)
* Whether the IT is to service business-focused information systems or project-based information systems (or both)
* Whether the IT investment is to provide knowledge work systems tools.

A distinction must be made between the resources and capabilities of SME and larger organisations. SMEs do not usually have individuals within the organisation with the necessary IT skills, knowledge or vision to identify IT/IS opportunities within their firm, and neither do they have the capacity to service and maintain their own IT infrastructures once installed. Furthermore, it is these firms that are least likely to be able to absorb the consequences of any IT investment failure, and they do not usually have the resources or the inclination to employ the requisite professional IT/IS advice upon a consultancy basis. Given that the vast majority of firms in the built environment fall into the SME category, the advice in this section of the chapter is directed towards their needs. There are two basic categories of information systems in the built environment:

1. Those that are focused upon the firm's business objectives (e.g. to acquire more clients or increase fee income)
2. Those that are targeted at the project or client's objectives (e.g. to complete the building on time, to cost, and to the expected level of quality).

Project based information systems are geared to enabling collaborative working practices and establishing project databases. IT is required to support both types of information system. A third role for IT in the built environment is to provide knowledge work systems for knowledge workers, as identified earlier in this chapter. This advice focuses upon the use of IT to support business objectives via company-based information systems or knowledge work systems.

Experience proves that *successful* IT investments have the following characteristics:

- IT investment is linked to core business objectives
- Successful IT strategies are driven from the top down in organisations, and should be championed by a senior partner/ manager
- IT investments should be linked to an IT strategy; that strategy should cover a period of 4–5 years and itself be subject to periodic review
- Users need to be involved in the selection and introduction of any new IT system to ensure that they have a sense of ownership of the system
- Successful introduction and acceptance of any new system must be supported by adequate and frequent staff training
- IT systems require effective user support and maintenance.

Other experience has shown that *unsuccessful* IT investments are characterised by:

- The lacking of integration with other IS/IT systems, thereby creating islands of computing
- Being technology driven (technology for technology's sake)
- Inadequate provision of IT equipment (e.g. the sharing of computers and software by users).

IT investment best practice

The methodology shown in Figure 6.3 is recommended to SMEs with regard to IT investment.

First, three tasks need to be undertaken concurrently; two if not all three of these can be performed internally with some guidance. The tasks are:

1. To survey existing IT facilities
2. To survey existing information systems
3. To identify the firm's business objectives.

The survey of the firm's existing IT infrastructure identifies the investments that have already been made, the outline specification of the equipment, the users, and the tasks performed. The IT survey should also include the definition of the firm's existing IT policy, which should cover very basic factors such as file-naming conventions, standard directory structures, and

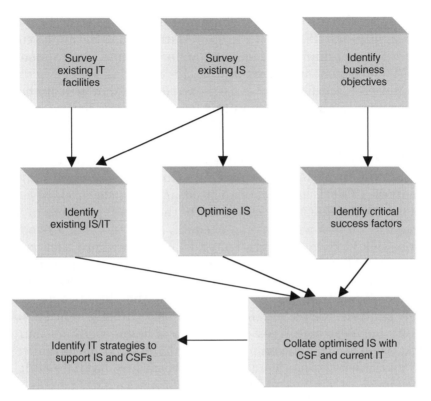

Figure 6.3 IT development methodology.

staff acceptable use policy. An acceptable use policy sets down rules and regulations regarding how employees may or may not use the information technology tools provided in the workplace. The intention is to avoid, and hopefully prevent, any malfunction of the company's supporting information technology due to the behaviour of employees. The survey should also identify the firm's back-up policy and procedures.

The objective of the IS survey is to identify how information flows through the firm and where and how it interfaces with the IT currently in use. This is the most difficult of the three initial tasks for the layman to undertake. It would normally be performed by a systems analyst; however, with a little background reading and some simplification, it is possible for this task to be performed in-house. The approach is to identify the tasks performed and where the data/information required to complete each task is created, stored, updated and deleted. The survey should also identify where and how the existing IT interfaces with the IS. The IS survey data should be analysed to identify core, duplicate and redundant systems, and to identify IS/IT development opportunities. Ideally 'clusters' of activity should be identified, and it is these clusters that will probably determine the best opportunities for the development of new IS – and equally, the existing ISs that are little used or of dubious value.

The firm's business objectives would normally be contained within the firm's business plan. A business threats/opportunities analysis (SWOT analysis) can used as a vehicle to clarify and focus the business objectives with the firm's partners if the business plan is weak or non-existent. All of the business objectives should then be investigated to identify where they relate to the IS/IT as defined in the IS and IT surveys. To complete this part of the exercise, critical success factors (CSFs) have to be developed from the business plan or the threats/opportunities analysis. CSFs are factors that are directly related to a firm's business objectives, are capable of being easily measured, and are used to determine whether the desired performance is being achieved and to identify where failure would result in serious shortcomings in business performance. Often CSFs will be both IS- and IT-related.

Examples of CSFs are:

- Improve the reliability, security and safety of computerised information
- Provide the means for partners to assess and compile fee bids from historical company records.

The results of the IT/IS surveys and the CSFs are then correlated to identify and prioritise IT opportunities. It is the IS requirements of the business objectives and/or the firm's IS system that should drive the identification of the IT facilities required. A number of alternative IT strategies should then be developed and subjected to discussion by the firm's partners. An IT strategy is simply a prioritised shopping list that is costed and linked to a time scale, normally a period of 4–5 years. Once an agreed strategy is developed, all investments should be made in accordance with that strategy. The strategy is usually subject to annual review but not wholesale change within the period of its life.

There are a number of related management issues that should be borne in mind at this stage:

- The acquisition of new technology should not lead to an associated increase in the firm's non-productive overhead costs. These can be minimised by adopting proven simple technology that is easily managed and maintained. Consideration should be given to acquiring a maintenance contract with a local IT consultancy for maintenance and repair of the IT systems.
- When purchasing equipment, seek turnkey contracts and quotations on a performance specification basis. Most SMEs can define what they want the system to do but are unable to specify or assess adequately technical tenders to support that requirement. Turnkey contracts also place the onus on the contractor to leave a fully working system in place upon completion of the contract.
- Actively seek out hardware suppliers who provide extended on-site warranties with their equipment. Three-year on-site warranties are now commonplace.
- Ensure that any investments made in software are with firms that have a good track record of continuous research and development, and are likely to be in business for some years to come.
- Be aware that even off the shelf software requires customisation for use in the workplace.

Conclusion

New technology should be seen by quantity surveyors as an opportunity rather than a threat. However, any investment needs to be based upon sound business practices and should always be related to the firm's business objectives. Furthermore, investments should be related to an IT strategy and championed by a senior member of the organisation. It is unlikely that new technology will provide a firm with any lasting commercial advantage, but failure to invest could lead to the firm being unable to compete with its rivals. Surveyors will be required to adopt and use new technology tools that enable collaborative working on a project basis, and it is quite likely that these systems will be based upon Internet technologies. To gain the

maximum benefit from any of these new technologies, surveyors
are going to have to adopt new working practices and probably
also develop new skills.

There follows a brief description of a contract administration
information system, developed and used by international
property consultants Franklin + Andrews, that demonstrates
the application of new technology to the commercial management of the provision of built assets.

Information systems in contract administration

One of the challenges facing organisations involved in the
creation and/or development of construction assets with a
variety of contracting partners, in widely varying commercial
climates, is the development and maintenance of standards – i.e.
standard data and standard methodologies. One of the
inevitable consequences of facing these challenges is the need
felt by such organisations for systems to support and reinforce
their understanding and control of the commercial process. The
normal response to this requirement is the development of a
database to support commercial management systems, but key
to the success of such an undertaking is a commercial management system focused exclusively on supporting the company's
delivery strategy.

In adopting such a system, there are a number of features
that are of critical importance. To be effective the system must
enable:

- The import or development and maintenance of a standard
  common estimating database
- The execution of estimates at any stage in the development
  of a project
- The execution and management of the tender and evaluation
  processes
- Commercial administration, including the measurement of
  progress and the certification of interim and final reports.

Standard database

Figure 6.4 illustrates the processes that take place during the
commercial cycle of a project from first estimates through to

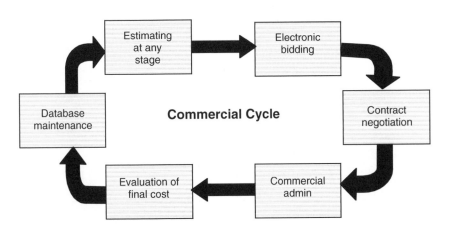

Figure 6.4 The commercial cycle.

calculation of final costs which are in turn supported by a variety of information contained in a standard cost data base. Two of the most significant problems associated with the development of a cost database are the cost of its maintenance and its responsiveness to changes in the business process. The selected system must address both these issues by adopting a cycle of cost information that ensures that the same data is used at every stage of the commercial process and is updated in real time at every tender, by using market information. The range of uses to which this data can be put will be determined by the method used to process the data once collected. Relatively straightforward exercises such as insurance valuations or approximate estimates for projects that are closely similar to existing assets can be completed using fairly simple databases, but higher risk exercises such as pre-tender estimating or target cost development need a more sophisticated and flexible platform.

Estimates at any level or stage

Careful structuring of the data allows its use in preparing estimates at the level of detail most appropriate to its purpose, from a feasibility study to a detailed pre-bid estimate, from interim valuation to final payment, as illustrated in Figure 6.5. By following this approach, an audit trail that will develop

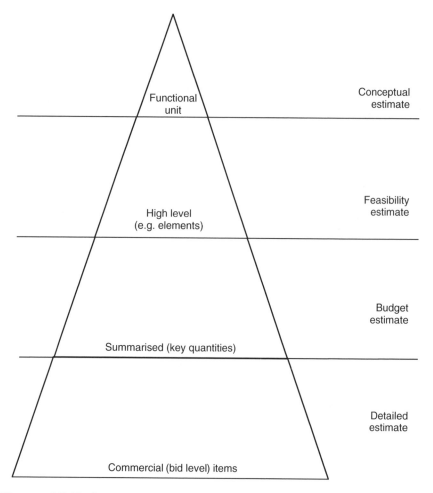

Figure 6.5 Estimates at any level.

through the life of the project is initiated at creation of the project, ultimately leading to a handover model that mirrors the company's asset register.

Integration of the tools available to the project is a vital feature. By integrating the commercial administration system with the planning system so that they share a common work breakdown structure, detailed cashflow forecasts and activity-based estimates that respond immediately to programme changes can be produced.

Although it is accepted that care is taken in the construction and maintenance of the database, any estimate should normally be subjected to a risk modelling exercise to identify risk, reduce it where possible, and provide a framework within which the remaining risk can be managed. This is normally achieved by means of data exchange with one of the many specialist risk-management packages currently available.

Tender evaluation process

Apart from removing the paper from the process, electronic bidding enables the contractor to prepare bid submissions electronically, carrying out the same sort of 'what-if' exercises as the client and collaborating with subcontractors to produce 'back-to-back' arrangements. The contractor is provided with all the tools necessary to prepare a bid by e-mail, on CD-ROM or, infrastructure permitting, online. This functionality is already in place and has been widely used in many areas of the world for several years. Full integration between contractor and client systems is less common, but, in order to reduce costs and the risk of error, is becoming more important. Bid evaluation and sensitivity testing facilities are vital to enable clients to safeguard themselves and fully comprehend the pricing strategy of the tenders. By using an electronic bidding system, bids can be compared against a pre-bid estimate based on the company's database, or against the other bids. The detailed understanding of the bid that the electronic bidding process provides is fundamentally important to the contract negotiation in an alliancing relationship. On completion of the evaluation and any subsequent contract negotiations, the contract can be awarded electronically and a control estimate produced. The control estimate is the benchmark against which progress, milestone payments or KPIs (discussed in Chapter 1) can be measured.

Commercial administration

As new parties take up their roles in the developing project, it is important that administration systems accommodate them in a collaborative environment. This might be under the umbrella of an alliancing contract or simply a traditional contract that seeks to avoid the duplications of effort traditionally associated with such methods. The system must not only permit access by all members of the team so that 'man-for-man marking' can be eliminated, but also provide either the functionality they require or the ability to integrate with their own systems. The shared facility enables work to be defined and evaluated, the calculation of invoice values, the evaluation of changes, and the maintenance of all necessary records for audit purposes. Ultimately, the value held in the system represents the final contract cost. Depending on the structuring of the data, it should be possible to tie all costs back to individual assets for rating, capital allowances, insurance evaluation or general ledger purposes. Beyond the individual project applications, integration with corporate financial systems is possible. By providing interfaces between the system and the client's financial systems, construction and maintenance costs can be included with the other elements of the firm's operations, and payments to contractors and suppliers can be made as quickly and cost effectively as possible.

The database

At the heart of the Franklin + Andrews system is the database. The size of the sample from which the database is created and the methodology adopted for its maintenance are of vital importance if it is to be used as the basis for estimating.

Database structure

The master database should consist of the cost information from a number of projects, combined to generate a standard. This is best illustrated by the example in Figure 6.6, which shows the arrangement of one level of a database holding five sample projects. The calculated rate might be generated by applying a formula to the sampled rates. This could be as simple as a straightforward mean, but is more likely to be based on the mode. Each new project's data would be added to the database and the average recalculated. In order to maintain the database, 'policies' or rules should be provided to govern 'retirement' or

Item	Set 1	Set 2	Set 3	Set 4	Set 5	Standard
Work Description	a1	a2	a3	a4	a5	*Calc Rate*

Figure 6.6. Database structure.

deletion of samples, as well as the methodology for dealing with particularly small samples. Having selected the statistical approach to database consolidation, much of the process can be handled automatically without losing any of the project specific detail.

In addition to the sample projects, a reference data set based on labour, plant and materials can be included as a 'project'. This can be mixed with the real data using the electronic equivalent of a mixer valve, which enables its significance in the calculation to be varied from 0 to 100%. This enables the preparation of data for work for which there is no precedent, or to condition particularly small samples. The above examples deal with only one level of data, but the required model will include a number of levels, perhaps up to nine, linked in one of two ways:

1. The first method of forming these links is to collect from the lower level all the gross values that need to be combined, and divide the value by a selected 'key quantity' to generate a higher level unit rate. For example, to generate a relationship between an item at the higher level for 'Concrete in foundation per cubic metre operational capacity', the items in the lower level relating to concrete would be collected and their total cost divided by the operational capacity to which they relate. This would generate a higher-level cost by cubic metre of product capacity. The results are recorded algebraically so that changes in the values at the lower level are automatically reflected at the higher level.

2. The second method is to define at the higher level the item to be linked, and then use the data at the lower level to evaluate one unit. For example, to generate a relationship between an item at the higher level for 'Concrete in bund walls' per linear metre, all the items necessary to construct a metre of typical bund wall would be measured and the total cost assigned to the higher level. The results are held algebraically so that changes in the values at the lower level are automatically reflected at the higher level.

Benchmarking

Unfortunately, any major employer of specialist construction skills can become the target of 'loading' of tenders. In the absence of published information, it is difficult to identify and therefore eradicate any factor that is applied specifically to the client. A clear solution to this problem is periodic benchmarking.

Interfacing with other systems

The commercial management system must be at the heart of the engineering business, alongside planning and financial systems. It goes without saying that the closer the integration of the systems, the more effective they will be. In practice, the security of the planning and financial functions normally limits the extent of the integration.

Further reading

Betts *et al.* (1999). *Strategic Management of IT in Construction.* Blackwell Science Ltd.
Castle, G. (2000). *Planned IT Infrastructure: Napier Blakeley Winter.* Building Centre Trust.

Web sites

BRE Centre for Construction IT; Best Practice for the use of IT in Construction: http://helios.bre.co.uk/iqit/
Construction Industry Computing Association (CICA); Making IT Work for Your Business, An Executive Briefing: http://www.cica.org.uk/sen_managers_it_briefing.pdf
IT Construction Best Practice; IT Applications: http://www.itcbp.org.uk/itcbp/ITapplications.jsp

7

Global markets – making ends meet

Introduction

This chapter will examine the role to be played by the quantity surveyor in the global marketplace, together with the problems faced by an organisation wishing to expand into European and/or global markets. Also included is an explanation of existing and proposed EU public procurement legislation and the procedures that must be followed to comply with European public procurement policy.

The multi-cultural team

The *Le Monde* cartoon featured in Figure 7.1 illustrates most quantity surveyors' perception or even experience of working in or with multi-cultural teams. Quantity surveyors have proved themselves to be adept in a diverse range of skills, often over and above their technical knowledge, with which they serve the needs of their clients. However, when operating in an international environment these skills and requirements are complicated by the added dimension of a whole series of other factors, including perhaps the most influential – cultural diversity. Companies operating at an international level in many sectors have come to realise the importance of a good understanding of cultural issues and the impact that they have on their business operations. However, in an increasingly global business environment in which the RICS is constantly promoting the surveyor as a global player (for example, in autumn 2001 the Institution launched the RICS Global Manifesto in the form of a consultation document), it is a fact that the realisation of the

Figure 7.1 Making ends meet (source: Serguei, *Le Monde*).

importance and influence of cultural diversity is still lacking in
many organisations seeking to expand their business outside of
the UK. Why were so few UK practices involved with the
rebuilding of Berlin, and why have so many UK practices failed
to make an impact in penetrating the French market? Figures
produced by the FIEC show that in 1999, amongst the EU
states, France and Sweden exported the largest percentages of
national construction turnover to world markets (Figure 7.2).

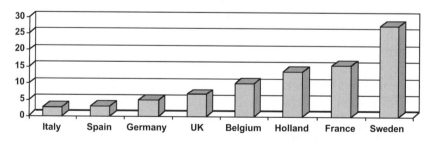

Figure 7.2 Construction exports in terms of volume of new overseas
contracts expressed as a percentage of national construction
turnovers in 1999 (source: FIEC-EIC; percentages indicative only).

What is it that makes these two countries particularly success-ful? Is it their approach to the construction process or their multi-cultural attitude that enables them to transcend national boundaries?

Today, international work is no longer separated from the mainstream surveying activity; EU Procurement Directives, GATT/GPA (Government Procurement Agreement of the World Trade Organisation) etc. are bringing an international dimen-sion to the work of the property professionals. Consultants from the UK are increasingly looking to newer overseas markets such as Europe and regions where they have few traditional histori-cal connections, such as South East Asia. Consultants must compete with local firms in all aspects of their services, includ-ing business etiquette, market knowledge, fees and, above all, delivering added value, in order to succeed. With the creation of the single European market in 1993, many UK firms were surprised that French and Spanish clients, for example, were not at all interested in the novelty value of using UK profes-sionals, and instead continued to award work on the basis of best value for money. As pointed out in Chapter 1, the UK construction industry and its associated professions have ploughed a lonely furrow for the last 150 years or so as far as the status and nature of the professions, procurement and approach to design are concerned, and it could be argued that this baggage makes it even more difficult to align with and/or adapt to overseas markets. Certainly companies like the French giant Bouygues, with its multidisciplinary *bureau d'études techniques*, have a major advantage in international markets because of their long established in-house capability to re-engineer initial designs and present alternative technical offers. Unlike in the UK, contractors in France are the largest employ-ers of construction professionals, which provides them with the capability to analyse design/construction problems in order to arrive at the best value added solution before submitting a bid to the client. Outside Europe, the USA, for example, has seen a decline in the performance of the US construction industry in international markets – a trend that has been attributed in part to its parochial nature in an increasingly global market. Inter-nally, strong trades unions exercise a vice-like grip on the Ameri-can construction industry, to the extent that in some states access to construction sites, even for casual visitors, is forbid-den. The same trades unions regularly identify organisations that use non-union labour by parking a giant, portable, inflatable 'black rat' outside their headquarters – behaviour not seen in the UK since the 1980s.

The construction industry, in common with many other major business sectors, has been dramatically affected by market globalisation. Previous chapters have described the impact of the digital economy on working practices; multinational clients such as Coca Coal and BP demand global solutions to their building needs, and professional practices as well as contractors are forging international alliances (either temporary or permanent) in order to meet the demand. It is a fast-moving and highly competitive market, where big is beautiful and response time is all important. The demands placed on professional consultants with a global presence are high, particularly in handling unfamiliar local culture, planning regimes and procurement practice, but the reward is greater consistency of workload for consultants and contractors alike. In an increasingly competitive environment, the companies that are operating at an international level in many sectors have come to realise that a good awareness and understanding of cultural issues is essential to their international business performance. Closeness and inter-relationships within the international business community are hard to penetrate without acceptance as an insider, which can only be achieved with cultural and social understanding. In order to maintain market share, quantity surveyors need to tailor their marketing strategies to take account of the different national cultures. Although some differences turn out to be ephemeral, when exploring international markets there is often a tightrope that has to be walked between an exaggerated respect, which can appear insulting, and a crass insensitivity, which is even more damaging. It comes as no surprise that cultural diversity has been identified as the single greatest barrier to business success.

It is no coincidence that the global explosion happened just as the Internet revolution arrived, with its 365 days a year/24 hours a day culture, allowing round the clock working and creating a market requiring international expertise backed by local knowledge and innovative management systems. Although it could be agued that in an e-commerce age cultural differences are likely to decrease in significance, they are in fact still very important, and remain major barriers to the globalisation of e-commerce. These differences extend also to commercial practice. Consider, for example, an issue that is crucial to e-commerce – privacy. The USA and Europe have traditionally had very different attitudes towards privacy, and Europe has adopted a very different approach to this problem *vis-à-vis* e-commerce. In the USA the approach towards making e-commerce a secure environment has been one of industry self-regulation, whilst the EU has decided that legislation is more appropriate and effective, as manifested by the European Directives on e-commerce and electronic signatures. Even in a digital economy an organisation still needs to discover and analyse a client's values and preferences, and there is still a role for trading intermediaries such as banks, trading companies, international supply chain managers, chambers of commerce, etc. in helping to bridge differences in culture, language and commercial practice. In an era of global markets purists could perhaps say that splitting markets into European and global sectors is a contradiction. However, Europe does have its own unique features, not least its public procurement directives, physical link and proximity to the UK, and for some states a single currency. Therefore, this chapter will first consider the European market, before looking at other opportunities.

Europe

During the early 1990s Euromania broke out in the UK construction industry, and 1 January 1993 was to herald the dawn of new opportunity. It was the day the remaining physical, technical and trade barriers were removed across Europe, and from now on Europe and its markets lay at the UK construction industry's feet. Optimism was high within the UK construction industry – after all, it seemed as though a barrier-free Europe with a multi-billion pound construction related output (825 billion euros in 2001, according to the European Construction Industry Federation; see Table 7.1) was the solution to falling turnover in the UK. Almost every month conferences were held on the theme of how to exploit construction industry opportunities in Europe. So, a decade later, has the promise been turned into a reality? There follows a review of developments in Europe, with an examination of the procurement opportunities for both the public and private sectors.

Table 7.1 EU construction-related output – 2001 (source: FIEC)

	Billion euros	*Percentage*
Germany	240.0	29.1
United Kingdom	113.4	13.7
France	98.8	12.0
Italy	91.6	11.1
Spain	87.4	10.6
Netherlands	40.6	4.9
Belgium	28.1	3.4
Austria	27.3	3.3
Sweden	24.5	3.0
Portugal	21.7	2.6
Denmark	19.5	2.4
Ireland	17.8	2.2
Finland	14.5	1.8
Total	**825.2**	**100**

European public procurement – an overview

Procurement in the European public sector involves governments, utilities (i.e. entities operating in the water, energy,
transport and telecommunications sectors) and local authorities
purchasing goods, services and works over a wide range of
market sectors, of which construction is a major part. For the
purposes of legislation, public bodies are divided into three
classes:

1. Central government and related bodies, e.g. NHS Trusts
2. Other public bodies, e.g. local authorities, universities etc.
3. Public utilities, e.g. water, electricity, gas, rail.

But why should quantity surveyors be interested? First, as a
procurement professional the quantity surveyor has much to
bring to the European procurement arena, not only for the existing Member States but also for the new states (including
countries such as Bulgaria and Estonia) that are predicted to
join the EU during the coming years. Many EU and prospective
EU States, and particularly but not exclusively those from the
former Eastern Block countries, still employ procurement
systems that favour the select few and preclude many, with all
the consequences that this brings for value for money. Secondly,

the widespread adoption of public private partnership throughout Europe as the preferred procurement route for a whole range
of projects, from the high profile TENs or Trans European
Networks to the provision of local healthcare facilities, means
that inevitably the quantity surveyor, whether in the public or
private sector, will increasingly require a working knowledge of
the rules governing the regulation of European public procurement. This is particularly so because there is increasing
evidence that failure by a contracting authority to comply with
EU Directives will be severely punished, as was demonstrated
in the so-called 'Harmon Case' (discussed later in this chapter),
potentially leading to a multi-million pound price tag for
damages. Finally, the sheer size and diverse nature of the
market, both existing and enlarged, make Europe a real and
exciting challenge.

The Directives – theory and practice

The EU Directives provide the legal framework for the matching of supply and demand in public procurement. A directive is
an instruction addressed to the EU Member States to achieve a
given legislative result by a given deadline. This is usually done
by transposing the terms of the directive into national legislation. Thus, the public works directives are instructions to the
Member States to modify their public procurement procedures
to comply with the requirements of the directive. In France this
was achieved by amending the *Code des Marchés Publics*, a code
that is part of national law, to take account of EU legislation.
In the UK, where there is no equivalent body of law governing
public procurement, the government sought to achieve implementation by means of the Public Works Contracts Regulations
1991 and by issuing instructions in the form of circulars to all
public purchasing authorities within the scope of the directives.
Alarmingly, statistics from Directorate General XV (DGXV), the
Brussels based directorate responsible for the regulation of the
internal market, reveal that one in eight of all internal market
directives had still to be implemented by Member States in
2000.

The European public procurement regulatory framework was
established by the public procurement Directives 93/36/EEC,
93/37/EEC and 92/50/EEC for supplies, works and services, and

Directive 93/38/EEC for utilities, which, together with the general principles enshrined in the Treaty of Rome (1957), established the following principles for cross-border trading (references apply to the Treaty of Rome):

- A ban on any discrimination on the grounds of nationality (Article 6).
- A ban on quantitative restrictions on imports and all measures having equivalent effect (Articles 30 to 36).
- The freedom of nationals of one Member State to establish themselves in another Member State (Articles 52 *et seq.*) and to provide services in another Member State (Articles 59 *et seq.*).

Enforcement Directives (89/665EEC and 92/13EEC) were added in 1991 in order to deal with breaches and infringements of the system by Member States.

Although adopted in the 1990s the Directives date back to the 1970s, and almost immediately it became apparent that they failed fully to reflect the changes resulting from the information technology revolution and the liberalisation of telecommunications across Europe that is enabling the expansion of e-procurement (see Chapter 5). Public procurement is quite different from private business transactions in several aspects; the procedures and practices are heavily regulated and, whilst private organisations can spend their own budgets more or less as they wish (with the agreement of their shareholders), public authorities receive their budgets from taxpayers and therefore have a responsibility to obtain value for money, usually based on cost. However, in recent years the clear blue water between private and public sectors has disappeared rapidly with the widespread adoption of public private partnerships and the privatisation of what were once publicly owned utilities or entities. The trend towards private involvement in public works has caused some difficulties within Directorate General XV, as for some time the legislation that empowers the system has been lagging behind developments in the market.

Unhindered cross-border tendering, free from tariffs, restrictions and protectionism, was the goal and the ongoing ideal of the European Commission and Directorate General XV in particular. The EU has legislated prolifically on public procurement since the 1985 White Paper *Completing the Internal Market*, which outlined the single market programme. The policy behind the creation of a single internal market was the belief that it could deliver greater economic performance and produce – according to the Cecchini report (1988), a 4.5–7% increase in the Community GDP. The volumes in monetary terms of goods and services that are procured within the states of the EU are truly immense. Contracts for public works and for the purchase of goods and services by public authorities and utilities account for around 14% of the Union's GDP. According to Internal Market Commissioner Frits Bolkestein, public procurement in the Union represents the equivalent of half of the German economy, some thousand billion euros per annum. Of this, approximately 200 billion euros per annum are generated by construction and allied services in the public sector, and should therefore be a focus of interest for the UK construction industry.

Given the historical background, it will come as no surprise that, more than 16 years after the introduction of European cross-border trading legislation, import penetration levels across all sectors are estimated to be only between 3 and 9%. The statistics at the more optimistic end of the range are produced by DGXV, by taking into account motor car manufacture, which involves components produced in several EU States that then cross borders to be assembled in other Member States. In addition, it is generally accepted that 85% of public authorities do not comply with the Directives in the knowledge that enforcement of compliance is virtually impossible. In 1996 the Commission published a Green Paper entitled *Public Procurement in the European Union: Exploring the Way Forward* as the basis for a dialogue about how to improve the system, in response to the following concerns of the Member States:

- That Member States were continuing to follow a buy national policy, in part because the Directives were failing to produce a transparent and equal system.
- That SMEs (small to medium-sized enterprises, i.e. organisations with a workforce of between 10 and 250) were being excluded.
- That the enforcement procedure to deal with breaches and infringements of the Directives was totally ineffective (see below).
- That the legal framework was too complex and unsuitable for the 'electronic age'.

Following receipt of nearly 300 responses, in 1998 the Commission published *Public Procurement in the European Union* with proposals for reform, namely:

*   Consolidation of the three classical sector Directives previously described into a single Directive for supply, works and services contracts.
*   A new directive for Utilities.
*   New rules on the use of competitive negotiated procedures to take account of public private partnerships and the PFI.
*   New rules to permit and control framework purchasing (this will be described later in the chapter).

In March 2001, after much debate and discussion, proposals were published to amend and consolidate the Directives in line with above. The implementation of the proposals is rather unpredictable, depending on the number and scale of disagreements within the European Parliament as well as between the Internal Market Council of Ministers and the Member States, but 2003–2004 is a realistically optimistic estimate, bearing in mind that DGXV has set the ambitious target that 25% of all procurement transactions should take place using electronic means by 2003.

Popular opinion is that Brussels is bristling with civil servants on very large salaries who do very little in return. In reality, DGXV badly lacks the resources to operate an effective public procurement regime and consequently has to adopt a reactive rather than proactive approach, particularly to questions of enforcement. Regrettably, the mechanisms for dealing with infringements to the Directives is to remain largely unchanged, and therefore it is almost inevitable that some Member States will continue to abuse the system. It is also to be regretted that the procurement Directives were not drawn up by procurement professionals but rather by politicians and civil servants, whose first priority was the promotion of cross-border competition; consequently good procurement practice was not high on the agenda. Many procurement professionals see this fundamental flaw as the principal reason for the limited success of the Directives. Another major concern (as mentioned previously) is the Commission's inability to enforce the directives and properly to punish contracting authorities who break the rules and cherry pick the parts of the law that they want to adhere to. Also regrettable is the lack of urgency shown by some Member States in complying with Directives and the manner in which compliance is achieved. For example, the Commission has instructed all Member States to put in place a system for dealing with contracting authorities who break the public procurement rules and award contracts unfairly. The UK has chosen to interpret the Directive by determining that the relevant forum for complaint within the UK shall be:

* the High Court in England and Wales
* the Court of Session in Scotland
* the High Court in Northern Ireland.

In practice this is referral to the existing legal system, with all the associated legal traps along the way. Whether this interpretation is in the spirit of the Directives is debatable. Many think that this is not an entirely appropriate method for dealing with disputes in what is considered to be a time sensitive environment; referral to these courts can take many months (or even years), by which time the project may well have been completed. Perhaps one of the most high profile infringement cases was the award of the contract to construct the Stade de France, at Saint Denis in the outskirts of Paris, for the 1998 Football World Cup Finals. Not only does this case involve the granting of a concession, an area that will be dealt with later in this chapter, but it also illustrates the woefully inadequate mechanisms for rectifying breaches in EU Directives. To understand the background to this case it is necessary to appreciate that in the mid-1990s France was in the midst of a deep recession, with unemployment levels of around 10% and an impending Presidential election. The construction of the Stade de France, like the World Cup Finals, was a high-profile project, the non-completion of which would have had disastrous consequences for the tournament, as well as widespread political and financial implications for many organisations. During 1996 a formal complaint was made to the Commission concerning the contract award procedure for the stadium. However, through a combination of time wasting and public sector bureaucracy it was not until March 1998 – 2 months before the first match was to be played and the stadium was completed – that the French Minister for European Affairs formally recognised the existence of the infringements. This admission of malpractice allowed a process that had been started nearly 2 years previously, and had included France being referred to the European Court of Justice by the then Single Market Commissioner Mario Monte, to be closed. However, no financial penalties were ever imposed, and the only outcome was a mild caution for the French contracting authority.

In a blatant foul, deserving of a red card, the French contracting authority permitted the winning French contractor to award a percentage of related construction contracts to local companies, in total breach of EU public procurement law. Equally importantly the contract was described in the contract announcement as a concession rather than a works contract, although the final contract had all operating aspects removed; this effectively gave the wrong impression to prospective bidders and excluded many non-French contractors who were misled by the Official Journal announcement.

However, despite criticisms of the mechanisms by which complaints against contracting authorities who breach the Directives are dealt with, recently, particularly in the UK, there have been a number of cases that have demonstrated for the first time that the UK remedies system does have teeth and that judges are prepared to take a robust attitude when it comes to dealing with breaches in EU procurement law. Most notably, in October 1999 in the High Court Judge Humphrey Lloyd QC delivered a decision in the case of *Harmon CFEM Facades (UK) Ltd v The Corporate Officer of the House of Commons* that attracted more interest than any procurement case so far before the UK courts. If this approach sets a precedent, contracting authorities have much more to fear from legal challenge than was thought to be the case. The case also has symbolic importance as the first major case in which an authority has been clearly condemned for a breach of the rules, and may now be open to significant damages liability as a result. The pity is that the judgment is nearly 300 pages long and took over a year to write – hardly a speedy and cost effective disputes resolution system! The case concerned a major £30 million contract for the fenestration work on the new office building for the House of Commons. The contract had been tendered under the Public Works Contract Regulations and the lowest bid, on a tender based on two options, was submitted by the French based Harmon CFEM Facades (UK) Ltd in the sum of £29.56 million. However, the contract was awarded to the more expensive UK company Seele/Alvis, who submitted a tender price of £32.26 million. The reason for selecting Seele/Alvis was given by the House of Commons as the alleged 'commercial' nature of the bid. His Honour Judge Humphrey Lloyd was not convinced, and accused the Commons of adopting a buy British policy in breach of the European Public Procurement rules. An interim payment of £1.9 million for loss of profit was made in June 2000, with hearings scheduled to decide the final damages, claimed by Harmon at £12 million. Like a curate's egg, EU public procurement is not all bad practice and procrastination. By contrast to the system adopted in the UK for dealing with disputes and breaches, the Danish system is much faster. In 1992 Denmark established *The Klagenoevnet for Udbud* – The Complaints Board. The board is a quasi-judicial tribunal, the members being drawn from technical experts who are able to deal quickly with alleged breaches in public procurement law and dispense remedies and/or financial penalties where appropriate. The speed of referral to the board means that, unlike in France or the UK, contracts unfairly awarded can be declared void and new bids sought without jeopardising the project.

The quantity surveyor and EU public procurement

So what of the current system – how is the quantity surveyor likely to come into contact with the European public procurement juggernaut, and what are the potential pitfalls? The following scenarios will be illustrated:

* A surveyor working within a public body (contracting authority) and dealing with a works contract.
* A surveyor in private practice wishing to bid for work in Europe as a result of a service contract announcement.

A surveyor within a public body

A quantity surveyor working within a body governed by public law (if in doubt, a list of European bodies and categories of bodies is listed in the Directives) should be familiar with procedures for compliance with European public procurement law. The Directives lay down thresholds above which it is mandatory to announce the contract particulars. The Official Journal is the required medium for contract announcements and is published five times each week, containing up to 1000 notices covering every imaginable contract required by central and local government and the utilities – from binoculars in Barcelona to project management in Porto. Major private sector companies also increasingly use the Official Journal for market research. In broad terms, the current thresholds for announcements in the Official Journal are:

1. For Works Contracts (i.e. construction), £3 611 395.
2. For Supplies and Service Contracts (i.e. quantity surveying, project management)
   - Central Government,   £93 896
   - Other public bodies,   £144 456
   - Utilities,   £288 912
   (NB: All figures exclude VAT).

However, DGXV actively encourages contracting authorities and entities to announce contracts that are below threshold limits.

Information on these impending tenders is published by the European Commission in Supplement S of the Official Journal of the European Communities, often otherwise known as the *OJ*, *OJEC* or *European Journal*. While the Official Journal used to be produced as a printed publication it is now only available on the Internet or on CD-ROM. The official European Commission website for tenders can be accessed at www.ted.eur-op.eu.int, and provides free access to the information, although it is not very user friendly. There are a number of commercial services that provide access to the same information and often include additional details, such as Tenders Direct, which can be accessed at www.tendersdirect.co.uk. Services such as Tenders Direct provide additional information and powerful easy-to-use search facilities that enable relevant tenders to be identified, as well as an e-mail alert service to provide notification of relevant tenders in the future (see Figure 7.3 and Appendix A).

Figure 7.3 shows the search results screen following a search for quantity surveying contracts. All the notices that match the search criteria are displayed in chronological order, with the most recent at the top of the list. The list includes the location and title of the project as well as the date of publication and, crucially, the deadline by which a supplier must have

Search Results _____

Your search for '*Quantity Surveying*' found **164** documents published in the last 6 months. *Click* the reference of the document you wish to view. [Note: If you wish to search for older documents please use the advanced search facility].

Page 1 2 3 4 5 6

Ref	Title	Published	Deadline
144110-2001	UK-Birmingham: project management, design, architectural, engineering, cost control and management s	31/10/01	29/11/01
144113-2001	UK-Glasgow: architectural, engineering, construction and related technical consultancy services	31/10/01	30/11/01
144133-2001	UK-London: building consultancy services	31/10/01	03/12/01
143454-2001	UK-Oxford: quantity surveying services	30/10/01	08/11/01
142416-2001	I-Milan: underground car park	27/10/01	17/12/01
141275-2001	UK-Stornoway: project-management services	25/10/01	19/11/01
141317-2001	UK-Oxford: architectural, engineering, construction and related technical consultancy services	25/10/01	07/11/01
140573-2001	UK-Epping: planning, design, project management and other related services	24/10/01	30/11/01
137785-2001	UK-Chelmsford: design services	18/10/01	09/11/01
137796-2001	IRL-Castlebar: design services	18/10/01	21/11/01
136991-2001	IRL-Dublin: architectural, engineering, construction and related technical consultancy services	17/10/01	
137147-2001	IRL-Tullamore: architectural, engineering and quantity surveying services	17/10/01	19/11/01
136442-2001	UK-Lewes: architectural, engineering and quantity surveying consultancy services	16/10/01	07/11/01
134999-2001	UK-Derby: advisory and information services	12/10/01	12/11/01
133625-2001	UK-Leeds: architectural, engineering and associated consultancy services	10/10/01	09/11/01
133653-2001	~~UK-Sheffield: quantity surveying and cost management services~~	~~10/10/01~~	~~15/10/01~~
132022-2001	B-Brussels: topographical services	06/10/01	16/11/01
129304-2001	~~UK-Lancaster: construction related professional services~~	~~29/09/01~~	~~29/10/01~~
128448-2001	~~IRL-Dublin: building design team services~~	~~28/09/01~~	~~30/10/01~~
128456-2001	~~UK-Arbroath: architectural design, engineering and related services~~	~~28/09/01~~	~~09/10/01~~
127734-2001	IRL-Tullamore: architectural, engineering and quantity surveying services	27/09/01	
127821-2001	~~UK-Ripley: building consultancy services~~	~~27/09/01~~	~~26/10/01~~
127858-2001	~~IRL- Castlebar: architectural design and associated services~~	~~27/09/01~~	~~31/10/01~~
127870-2001	IRL-Dublin: architectural, engineering and quantity surveying services	27/09/01	16/11/01
127161-2001	~~UK-Birmingham: architectural design, engineering, quantity surveying and related services~~	~~26/09/01~~	~~23/10/01~~
127168-2001	~~UK-Wolverhampton: architectural, civil engineering and related works and services~~	~~26/09/01~~	~~29/10/01~~
126563-2001	~~IRL- Castlebar: design services~~	~~25/09/01~~	~~31/10/01~~
125822-2001	~~UK-Worcester: architectural, engineering, project management and related services~~	~~22/09/01~~	~~22/10/01~~
124602-2001	~~UK-Shrewsbury: professional design and associated services~~	~~20/09/01~~	~~03/10/01~~
123216-2001	UK-Arbroath: architectural design, engineering and related services	18/09/01	

Page 1 2 3 4 5 6
Next Page

Figure 7.3 Tenders Direct search results.

confirmed their interest in tendering. An abstract of the notice can be viewed by clicking on its reference number, although to obtain the full tender notice users are required to register with Tenders Direct and pay a small fee for each notice they wish to download. It should be noted that the search results also show UK public procurement opportunities for quantity surveyors, and as such can be useful to UK-based organisations too.

As mentioned previously, the Directives categorise contracts as follows.

Public Works contracts are defined in Article 1 of the Directive as those contracts that:

1. Have as their object either the execution or both the execution and design of works related to one of the activities related in Annex II, or
2. Are the outcome of building or civil engineering works taken as a whole that are sufficient of themselves to fulfil an economic and technical function, or
3. Execute, by whatever means, work corresponding to the requirements specified by the contracting (public) authority.

Public Service contracts are defined in Article 1 of the Directive as, among other things:

• Contracts of pecuniary interest for employment, e.g. professional quantity surveying services.

A further category is becoming increasingly important, and that is concessions. Concessions to run public services are not defined by the Treaty of Rome and have caused DGXV a number of legal headaches in the form of complaints concerning infringements of community law when public authorities have called on economic operators' know-how and capital to carry out complex operations. The principal problem stems from the way that concession contracts have proliferated in recent years with the increased use of public private partnership projects across Europe. In an attempt to get to grips with the situation, a Draft Commission Interpretative Communication on Concessions was issued by the Commission in February 1999. and subsequently, in April 2000, after consultations with interested parties, the Communication was adopted with some minor alterations. In this document the Commission has defined concessions as 'cases where public authorities entrust a third party to totally or partially manage infrastructure projects or other public services on its behalf and for which the third party assumes the operating risks'. A concession can be a PPP form, similar to a design, build, operate and finance contract, except that the private sector contractor retains some operational risk and recovers its costs either through direct user charges or through a mixture of user charging and public subvention. In this context it is important to note that PPP concessions will not always meet the strict definition of concessions under EU procurement directives, as compliance will depend on the level of exploitation of the asset. Note also that it is only works concession contracts with a value greater than the threshold of 5 million euros that must comply; service concessions, where the threshold for compliance is much lower, are not regulated by the Directives. The question of which procurement Directive applies is only significant in the context of concession contracts. For example:

- In roads contracts, where there is a hard toll or a combination of hard and soft tolling, providing that the hard tolling element is such that there is an exploitation risk remaining with the contractor; for example, where toll charges are fixed by the public sector sponsor at an artificially low level, which does not cover exploitation costs, then the contractor may receive a shadow tolling subsidy.
- In water contracts, where there is a user charge that results in an exploitation risk remaining with the contractor.

It should be noted that an interpretive communication is not legally binding in itself, as it does not bring anything new to the legal framework. It is rather a clarification of the Commission's understanding of a legal text, which may be useful for the operators concerned. Once the nature of the project has been established, the contracting authority can begin the procedure for announcing the details to prospective bidders.

The announcement procedure involves three stages:

1. Prior information notices (PIN) or indicative notices
2. Contract notices
3. Contract award notices (CANs).

Examples of these notices can be found in Annex IV of the Directive, and examples of all three types of notice are given in the Appendix (see pp. 278–81).

A prior information notice, or PIN, that is not mandatory, is an indication of the essential characteristics of a works contract and the estimated value. It should be confined to a brief statement, and posted as soon as planning permission has been granted. The aim is to enable contractors to schedule their work better and allow contractors from other Member States the time to compete on an equal footing. Where work is subdivided into several lots, each one the subject of a contract, the aggregate value should be taken into account when determining whether the threshold has been exceeded. For example, a contract for the construction of a new prison has been divided into three lots, estimated at 3 million euros, 1.2 million euros and 1.6 million euros respectively. The estimated value for procurement purposes is therefore 5.8 million euros, and the Directives will apply. Note that had one of the work packages been valued at less than 1 million euros, the authority would have been allowed to awarded it freely without reference to the Directives; however, it would still have been included within the total value for calculating thresholds – that is, 3 million plus 1.2 million plus 900 000 euros, total 5.1 million euros, with the Directives applying to work packages one and two only. It is a common complaint that authorities split contracts to avoid the Directives; although specifically prohibited by the Directives this action is difficult to prove, but it does help to promote a buy national policy by States. The prior information notice can be a useful market-testing tool in the case of public private partnerships, as it affords the contracting authority the opportunity to assess the potential interest from consortia as well as the financial viability and business case of the project that is being proposed. Increasingly the commission is encouraging public bodies to post a purchaser profile, which is a statement of the kind of supplies, services and works contracts that a particular body is likely to require, and at what intervals and quantities.

Contract notices are mandatory and must include the award criteria, which can be based on either the lowest price or the most economically advantageous tender, specifying the factors that will be taken into consideration.

Once drafted, the notices are published, five times a week, via the Publications Office of the European Commission in

Luxembourg in the Supplement S to the Official Journal via the Tenders Electronic Daily (TED) database, and translated into the official languages of the community, all costs being borne by the Community. TED is updated twice weekly and may be accessed through the Commission's web site at http://simap.eu.int. Extracts from TED are also published weekly in the trade press.

In order to give all potential contractors a chance to tender for a contract, the Directives lay down minimum periods of time to be allowed at various stages of the procedure – for example, in the case of Open Procedure this ranges from 36 to 52 days from the date of dispatch of the notice for publication in the Official Journal. Restricted and negotiated procedures have their own time limits. These timescales should be greatly reduced with the wide-scale adoption of electronic procurement.

The production of tender documents for a construction project for the domestic market is in itself a difficult task that requires a good deal of experience and professional know-how. To produce documentation suitable for transmission to all EU Member States in a form that maintains transparency and accuracy demands due diligence of the highest order. To aid contracting authorities in accurately describing proposed works, the Common Procurement Vocabulary (CPV) has been developed to facilitate fast and accurate translation of contract notices for publication in the Official Journal. CPV is a series of nine digit codes that relate to the area of activity of goods, services or works for which tenders are invited; at present the use of CPV in contract notices is optional. Certainly CPV makes searching for particular market sectors on databases much easier than has previously been the case. Although the use of CPV is currently not mandatory, if the contracting entity does not use the classification system the Commission will translate it, with perhaps unusual if not entirely unpredictable results. For example, the term 'Christmas trees' is widely used within the oil and gas industries as vernacular for blow-out valves. Consequently, when Christmas trees were referred to by an entity in oil and gas but not classified with CPV in the tender notice, the commission assumed that the entity was attempting to procure trees and classified the notice accordingly! However, in August 2001 the European Commission adopted a proposal that will establish CPV as the only system used for the classification of public procurement announcements. The motive behind the decision was to ensure that the subject matter of contracts could be accurately identified, allowing automatic translation of tender notices into all official Community languages.

Contract award notices inform contractors about the outcome of the procedure. If the lowest price was the stated criterion, this is not difficult to apply. If, however, the award was based on the 'most economically advantageous tender', then further clarification is required to explain the criteria – e.g. price, period for completion, running costs, profitability and technical merit, listed in descending order of importance. Once established, the criteria should be stated in the Contract Notices or contract documents.

Award procedures

The surveyor must decide at an early stage which award procedure should be adopted. The choices are as follows:

1. *Open procedure*, which allows all interested contractors to submit tenders.
2. *Restricted procedure*, which initially operates as the open procedure but then the contracting authority only invites certain contractors, based upon their standing and technical competence, to submit a tender. Under certain circumstances, for example extreme urgency, this procedure may be accelerated – an action that may be seen to restrict competition. Authorities who use this tactic to excess will be reprimanded by DGXV.
3. *Negotiated procedure*, in which the contracting authority negotiates directly with the contractor of its choice. The Directives allow this procedure to be utilised without the use of a prior information notice, and it may be subjected to accelerated procedures as described previously. The Directives lay down the cases in which this procedure may be followed, described collectively as cases where it is strictly necessary to cope with unforeseeable circumstances (events that overwhelmingly transcend the normal bounds of economic or social life, such as an earthquake or flood). Clearly, the UK Pimlico school project is an example where the Commission did not consider the circumstances right to use this procedure and disapproved of the manner in which the contract was awarded, as it was adjudged that negotiations had been used deliberately to avoid competitive tendering.

4. *Competitive dialogue procedure* (originally a sub-clause of the negotiated procedure), which is the Commission's attempt to deal with the increasing number of PPP/PFI contracts. However, this new pathway is thought by many to preclude rather than include PPP, as dialogue has to continue with all pre-qualified consortia and without the possibility of proceeding to the negotiation/preferred bidder stage (see Table 4.5). The proposal to introduce this new procedure in 2002–2003 for what the Commission refers to as 'complex' projects is one of the most controversial in recent times, and has aroused concern that it could damage PPP deals or even make the current UK model unworkable. The basis of this procedure is to accommodate contracts, such as the PFI and PPPs, where the contracting authority leaves it up to the bidders to determine the best and most suitable financial or technical solution – the so-called 'complex' contract. The main concern is that the procedure precludes the use of steps 9–11 in the UK model (see Table 4.5) and does not allow the contracting authority to discuss with preferred bidders the details of their submission. In effect this means that three preferred bidders would be required to work up and submit final proposals with the inevitable sharp increase in tender costs – a throwback to the bad old days of early PFI deals. The proposed procedure is as follows; consortia for a PPP contract are invited to submit, via the *OJ*, pre-qualification details, including outline solutions, for a new project. After submission, the public contracting authority can choose either to use one of the submissions as a basis for developing a detailed solution, or to pick and mix the submissions as a basis for developing a detailed solution. Thereafter not less than three consortia will be invited to develop their bids based on one of the above. After selection of a preferred consortium no further negotiation or dialogue will be permitted – the deal must stand. The UK government has made strong representations to the Commission that it cannot implement the directive as proposed. However, it does appear that the Commission will proceed with its intention.

In addition to the above award criteria, the Commission is increasingly engaged in the potential for interaction between public procurement and environmental and social matters. For example, organisations bidding for public projects might have to demonstrate the contribution that the project could make to the environment and social matters. Architects and/or engineers might be given clear instructions to design, for example, a low-energy consuming administrative building, taking into account not only insulation and the use of specific construction materials but also the installation of solar cells for the generation of warmth. If these sorts of considerations were to be selection criteria, then the additional question arises: will environmental and social criteria take over from the most economically advantageous as the basis of awarding a contract, or will bids be evaluated on a 50–50 basis? In July 2001 the Commission published an Interpretive Communication on the subject that referred to the possibility of taking whole life costs into consideration when awarding a contract.

Frameworks

Framework agreements are used increasingly in both public and private sectors. They are used in repetitive contracts and eliminate the need to call for tenders for similar types of services or materials. Multi-supplier frameworks follow two procedures: in the first, the supplier for each order is selected solely on the basis of the original bid; in the second, suppliers have the opportunity to amend or complete bids before the winner is selected for each order.

For example, in a *Type 1* agreement suppliers of building components are invited to submit prices for 200 standard items, together with delivery rates to different parts of the UK. Six suppliers are selected, based on price, and form the framework for supplying the items at the tender prices for a period of up to 4 years. In a *Type 2* agreement, surveyors submit rates and details of expertise, based on several specimen briefs. Three are selected based upon rates and expertise. A faxed brief is provided for each project, and firms indicate available consultants and current rates; a surveyor is selected based on rates and expertise of those available.

Electronic auctions

The Internet is making the use of electronic auctions increasingly more attractive as a means of obtaining bids in both public and private sectors; indeed it can be one of the most transparent methods of procurement. At present electronic auctions can be used in both open and restricted framework procedures. The system works as follows:

*   The framework (i.e. of the selected bidders) is drawn up.
*   The specification is prepared.
*   The public entity then establishes the lowest price award criterion, for example with a benchmark price as a starting point for bidding.
*   Reverse bidding on price then takes place, with framework organisations agreeing to bid openly against the benchmark price.
*   Prices/bids are posted up to a stated deadline.
*   All bidders see the final price.

Technical specifications

At the heart of all domestic procurement practice is compliance with the technical requirements of the contract documentation in order to produce a completed project that performs to the standards of the brief. The project must comply with national standards and be compatible with existing systems and technical performance. The task of achieving technical excellence becomes more difficult when there is the possibility of the works being carried out by a contractor who is unfamiliar with domestic conventions and is attempting to translate complex data into another language. It is therefore very important that standards and technical requirements are described in clear terms with regard to levels of quality, performance, safety, dimensions, testing, marking or labelling, inspection, and methods or techniques of construction, etc. References should be made to:

*   *A Standard*: a technical specification approved by a recognised standardising body for repeated and continuous application.
*   *A European Standard*: a standard approved by the European Committee for Standardisation (CEN).
*   *European technical approval*: a favourable technical assessment of the fitness for use of a product, issued by an approval body designated for the purpose (sector-specific information regarding European technical approval for building products is provided in Directive 89/106/EEC).

- *Common technical specification*: a technical specification laid down to ensure uniform application in all Member States, which has been published in the Official Journal.
- *Essential requirements*: requirements regarding safety, health and certain other aspects in the general interests that the construction works must meet.

Given the increased complexity of construction projects, the dissemination of accurate and comprehensive technical data is gaining in importance. It is therefore not surprising that the Commission is concerned that contracting authorities are, either deliberately or otherwise, including discriminatory requirements in contract documents. These include:

- Lack of reference to European standards.
- Application of technical specifications that give preference to domestic production.
- Requirements of tests and certification by a domestic laboratory.

The result of this is in direct contravention of Article 30 of the Treaty of Rome, and effectively restricts competition to domestic contractors.

A surveyor in private practice

Professional surveyors cannot help but be tempted by the calls for bids to be submitted to provide a whole range of services to contracting authorities through Europe (see Figure 7.3). This section of the chapter should not be read in isolation; the preceding section should first be studied in order to appreciate fully the general public procurement environment. Anecdotal information of unsuccessful and costly attempts to break into the European public procurement market abounds; many of the difficulties are attributable to lack of knowledge of the Directives and basic language competencies, as well as a large degree of naivety. It should be remembered that the European market is not a level playing field and that any attempt to break into new markets requires detailed market research, including the establishment of local contacts, for in this respect there is no difference between public and private sectors. When selecting Member States to target, research must be carried out to discover whether the

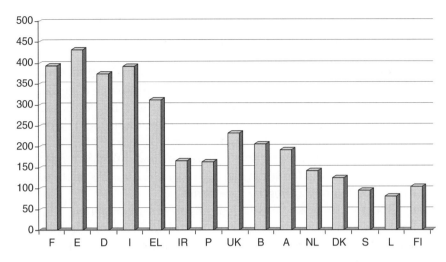

Figure 7.4 Infringements of EU Directives (source: Secretariat General, European Commission).

Directives have been implemented – after all, there is little point in trying to win work in a Member State that cherry-picks which parts of EU law it implements. In a review concluded in July 2001 by DGXV, the best performing countries were Denmark, Finland, Sweden and Spain, while Greece, France and Belgium were lagging behind with implementation deficits of more than twice those of the best performing countries. Another indicator of a State's willingness to comply with EU Directives is its track record of infringements. Figure 7.4 illustrates a breakdown, by Member State, of the 3414 cases of infringements to Directives under examination at 25 July 2001.

As indicated, Denmark once more fares well compared with France, Germany and the UK. Other positive indicators are the procedures that are in place for dealing with infringements – for example, is there a separate national review board for dealing quickly with complaints, as in Denmark?

Following the initial selection, there are other checks that the surveyor should make when carrying out a search of contract announcements:

- When searching TED for PIN or contract notices, look for announcements with open award procedures and a comprehensive specification; this is a good indication that the contracting authority is determined, at least on the face of it, to make the procurement process as transparent as possible.

* Avoid announcements with single- or multi-named contacts – it's far better to have a general administrative person as a contact for further information. Single-name points of contact could indicate that the named person is carrying out an unofficial pre-selection screening!

* Avoid announcements with lists of acceptable tenderers. Although widely condemned by the DGXV this still exists and is obviously a restrictive practice.

* Check the Contract Award Notices to discover which Member States seem to be awarding a high proportion of contracts to domestic contractors. This task can be carried out quite simply via TED.

* Check the countries that use negotiated or accelerated procedures to excess, as this procurement path effectively excludes all but domestic contractors.

What therefore is the most effective way for a surveyor to access information concerning European public procurement opportunities?

As described previously, there are several pathways to access information on current contracts, ranging from freely available services to subscription services such as Tenders Direct, and logging on to these services could be a good starting point. Familiarity with the Common Procurement Vocabulary, an example of which is given in Table 7.2, will save time when carrying out a search, particularly if TED is being used. The appropriate CPV inserted into code in the search engine of TED will retrieve all the information on the database associated with

Table 7.2 An example of common procurement vocabulary

74000000	Architectural, construction, legal, accountancy and business services		
	74200000	Architectural, engineering, construction and related technical consultancy services	
		74210000	Technical consultancy services
			74216000 Construction-related services

a particular market sector. This facility may also be used to analyse Tender Award Notices.

It is a fact of life that the larger the organisation the greater the resources, both physical and financial, that are available to take advantage of new markets and opportunities. One of the major concerns that emerged from the Public Procurement Green Paper was the way in which SMEs were being disadvantaged, a concern shared by DGXV, which, by a number of initiatives, has attempted to redress the balance of opportunity. One of the primary means of redressing the balance is the interface between procurement and information technologies. The Commission is promoting, through a series of Directives and other initiatives, the Internet as the preferred method of public procurement. The European Commission funded SIMAP project was launched in order to encourage best practice in the use of information technology for public procurement. The development of information technologies and the dramatic reduction in telecommunication costs have created favourable conditions for moving from paper-based commerce towards full e-commerce and addressing the whole procurement process, including bids, awards of contracts, delivery, invoicing.

The primary objectives of SIMAP are to improve the dissemination and quality of information on procurement opportunities and encourage electronic data interchange between purchasers and suppliers. It is thought that an electronic procurement system will reduce transaction costs and time, and improve the management of the system as a whole. The development of electronic procurement is a joint responsibility between the public and private sectors; the public authorities' role is to define the legal framework and to facilitate development of ICT tools, while the private sector must develop and implement applications and ensure technical interoperability.

Public procurement beyond Europe

There are no multilateral rules covering public procurement. As a result, governments are able to maintain procurement policies and practices that are trade distortive. That many governments wish to do so is understandable; government purchasing is used by many as a means of pursuing important policy objectives that have little to do with economics – social and industrial policy objectives rank high amongst these. The plurilateral Government Procurement Agreement (GPA) partially fills the void. GPA is based on the GATT provisions negotiated during the 1970s, and is reviewed and refined at meetings (or rounds) by ministers at regular intervals. Its main objective is to open up international procurement markets by applying the obligations of non-discrimination and transparency to the tendering procedures of government entities. It has been estimated that market opportunities for public procurement increased ten fold as a result of the GPA. The GPA's approach follows that of the European rules. The Agreement establishes a set of rules governing the procurement activities of member countries and provides for market access opportunities. It contains general provisions prohibiting discrimination as well as detailed award procedures. These are quite similar to those under the European regime, covering both works and other services involving, for example, competition, the use of formal tendering and enforcement, although the procedures are generally more flexible than under the European rules. However, GPA does have a number of shortcomings. First, and perhaps most significantly, its disciplines apply only to those World Trade Organisation members that have signed it. The net result is a continuing black hole in multilateral WTO rules that denies access or provides no legal guarantees of access to billions of dollars of market opportunities in both the goods and services sector. The present parties are the European Union, Aruba, Norway, Canada, Israel, Japan, Liechtenstein, South Korea, the USA, Switzerland and Singapore.

Developments in public procurement

As in the private sector, information technology is the driving force in bringing efficiency and added value to procurement. However, despite the many independent research projects that have been undertaken by the private sector, the findings cannot simply be lifted and incorporated into the public sector due to the numerous UK and European Community regulations that must be adhered to. Notwithstanding these potential problems, UK government has set a target that by the end of 2002 the majority of government tenders should be procured electronically, although this target has been called into question by many who regard it as wildly optimistic.

Of all the strands of the e-business revolution, it is e-procurement that has been the most broadly adopted, has laid claim to the greatest benefits and accounts for the vast majority of electronic trading. Despite the fact that the EU is legislating on the matter of electronic signatures and the present mandatory requirements for the use of written contracts, for example, a general policy forcing electronic communication cannot be centrally enforced but only implemented on the wish of the procuring entity. Nevertheless, a survey carried out on behalf of the EU in 2000 showed that, of the existing electronic procurement systems in use, building and construction was offered by all of them and was the top-ranked sector, with a usage rate of 72%.

The stated prime objective of electronic tendering systems is to provide central government with a system and service that replaces the traditional paper tendering exercise with a web-enabled system that delivers additional functionality and increased benefits to all parties involved with the tendering exercise.

The perceived benefits of electronic procurement are as follows:

- Efficient and effective electronic interfaces between suppliers and civil central governments, departments and agencies, leading to cost reductions and timesaving on both sides.
- Quick and accurate pre-qualification and evaluation, which enables automatic rejection of tenders that fail to meet stipulated 'must have criteria'.
- A reduced paper trail on tendering exercises, saving costs on both sides and improving audit.
- Increased compliance with EU Procurement Directives, and best practice procurement with the introduction of a less fragmented procurement process.
- A clear audit trail, demonstrating integrity.
- The provision of quality assurance information – e.g. the number of tenders issued, response rates and times.
- The opportunity to gain advantage from any future changes to the EU Procurement Directives.
- Quick and accurate evaluation of tenders.
- The opportunity to respond to any questions or points of clarification during the tender period.

- Reduction in the receipt, recording and distribution of tender submissions.
- Twenty-four-hour access.

Figure 7.5 illustrates the possible applications of e-procurement.

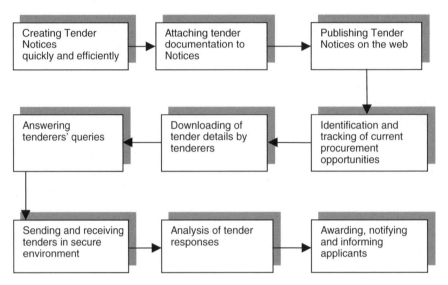

Figure 7.5. e-Procurement.

Europe and beyond

The effect of culture on surveyors operating in international markets

As discussed in the opening to this chapter, culture can be a major barrier to international success. Culture must first be defined and then analysed so that it can be managed effectively; thereafter there is the possibility of modelling the variables as an aid to business. A business culture does not change quickly, but the business environment from which it is derived and with which it constantly interacts is sometimes subject to radical and dramatic change. The business culture in a particular country grows partly out of what could be called the current business environment of that country. Yet business culture is a much broader concept, because alongside the impulses that are derived from the present

Table 7.3 The effects of culture on business conducted through European companies (source: Elucidate: The Report)

China	Cultural differences are as important as an understanding of Asian or indeed other foreign languages.
Far East	Crucial to know etiquette/hierarchical structure/manner of conduct in meetings.
Germany	Rigid approach to most operational procedures.
Middle East	Totally different culture – time, motivation, responsibility.
Russia	Inability to believe that the terms and conditions as stated really are what they are stated to be.
SE Asia	Strict etiquette of business in S. Korea and China can be a major problem if not understood.
France	Misunderstandings occurred through misinterpretation of cultural differences.

business environment there are the historical experiences of the business community. For example, as discussed in Chapter 1, the 1990 recession caused widespread hardship, particularly in the UK construction industry. There have been many forecasts of doom during the early twenty-first century from analysts drawing comparisons between the state of business at this time with that in 1990, when record output, rising prices and full employment were threatening to overheat the economy as well as construction – can there be many quantity surveying practices in the UK that are not looking over their shoulders to see if and when the next recession is coming? Table 7.3 outlines a sample of the responses by 1500 European companies questioned during a study into the effects of culture on business.

So what is culture? Of the many definitions of culture, the one that seems most accurately to sum up this complex topic is 'an historical emergent set of values'. The cultural differences within the property/construction sectors can be seen to operate at a number of levels, but can be categorised as follows:

1. Business/economic factors – e.g. differences in the economic and legal systems, labour markets, professional institutions etc. of different countries.
2. Anthropological factors, as explored by Hofstede (1984). The Hofstede IBM study involved 116 000 employees in 40 different countries, and is widely accepted as being the benchmark study in this field.

Of these two groups of factors, the first can be regarded as fairly mechanistic in nature, and the learning curve for most organisations can be comparatively steep. For example, the practice of quantity surveyors in France of paying the contractor a sum of money in advance of any works on site many seem anathema, but it is usual practice in a system where the contractor is a trusted member of the project team. It is the second category of cultural factors, the anthropological factors, that is more problematic. This is particularly so for small and medium enterprises, as larger organisations have sufficient experience (albeit via a local subsidiary) to navigate a path through the cultural maze.

Perhaps one of the most famous pieces of research on the effects of culture was carried out by Gert Hofstede for IBM. Hofstede identified four key value dimensions on which national culture differed (Figure 7.6), a fifth being identified and added by Bond in 1988 (Hofstede and Bond, 1988). These value dimensions were power distance, uncertainty avoidance, individualism/collectivism, and masculinity/femininity, plus the added long-/short-termism. Although neatly categorised and explained below, these values do of course in practice interweave and interact to varying degrees.

- *Power distance* indicates the extent to which a society accepts the unequal distribution of power in institutions and organisations, as characterised by organisations with high levels of hierarchy, supervisory control and centralised decision-making. For example, managers in Latin countries expect their position within the organisation to be revered and respected. For French managers the most important function is control, which is derived from the hierarchy.

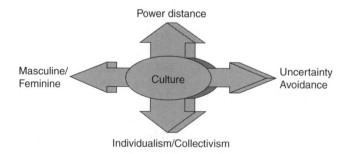

Figure 7.6 Hofstede's cultural determinants.

- *Uncertainty avoidance* refers to a society's ability to cope with unpredictability. Managers avoid taking risks and tend to have more of a role in planning and co-ordination. There is a tendency towards a greater quantity of written procedures and codes of conduct. In Germany managers tend to be specialists and stay longer in one job, and feel uncomfortable with any divergence between written procedures – for example the specification for concrete work and the works on site. They expect instructions to be carried out to the letter.
- *Individualism/collectivism* reflects the extent to which the members of a society prefer to take care of themselves and their immediate families as opposed to being dependent on groups or other collectives. In these societies, decisions would be taken by groups rather than individuals, and the role of the manager is as a facilitator of the team (e.g. Asian countries). In Japan tasks are assigned to groups rather than individuals, creating stronger links between individuals and the company.
- *Masculinity/femininity* refers to the bias towards an assertive, competitive, materialistic society (masculine) or the feminine values of nurturing and relationships. Masculine cultures are characterised by a management style that reflects the importance of producing profits, whereas in a feminine culture the role of the manager is to safeguard the wellbeing of the work force. To the American manager, a low head count is an essential part of business success and high profit; anyone thought to be surplus to requirements will be told to clear his/her desk and leave the company.

As a starting point for an organisation considering looking outside the UK for work, Figure 7.7 may be a somewhat light-hearted but useful discussion aid to help recognise and identify the different approaches to be found towards organisational behaviour in other countries/cultures – approaches that if not recognised can be a major roadblock to success.

Developing a strategy

The development process, when carried out internationally, is particularly complex to manage due to the weaving together of various cultures, including language (both generic and technical), professional standards and construction codes, design approaches

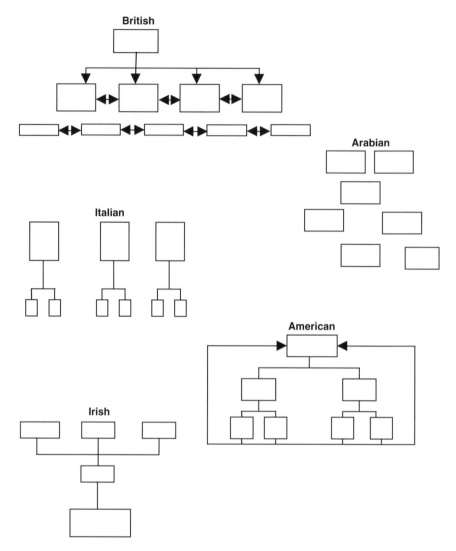

Figure 7.7. Organisational chart (source: adapted from *International Management*, Reed Business Publishing).

and technology, codes of conduct, and ethical standards. Technical competency and cultural integration must be taken as read.

The competencies necessary to achieve cultural fluency can therefore be said to be:

- Interpersonal skills
- Linguistic ability

- Motivation to work abroad
- Tolerance of uncertainty
- Flexibility
- Respect
- Cultural empathy.

Case studies of SMEs show that 60% of companies react to an approach from a company in another country to become involved in international working. The advantages of reacting to an enquiry are that this approach involves the minimum amount of risk and requires no investment in market research, but consequentially it never approaches the status of a core activity, is usually confined to occasional involvement and is only ever of superficial interest. However, to be successful the move into overseas markets requires commitment, investment and a good business plan linked to the core business of the organisation.

A traditional approach taken by many surveying practices operating in world markets, particularly where English is not the first language, is to take the view that the operation should be headed up by a native professional, based on the maxim that, for example, 'it takes an Italian to negotiate with an Italian'. Although recognising the importance of cultural diversity, the disadvantages of this approach are that the parent company can sometimes feel like a wallflower, there is no opportunity for parent company employees to build up management skills, and in the course of time local professionals may decide to start their own business and take the local client base with them. If culture is defined as shared values and beliefs, then no wonder so many UK companies take this approach. How long, for example, would it take for a British quantity surveyor to acquire the cultural values of Spain?

As a starting point, a practice considering expanding into new markets outside the UK should undertake the following.

1. Carry out extensive market research:
   - Ensure market research covers communication (language and cultural issues).
   - Make frequent visits to the market; it shows commitment rather than trying to pick up the occasional piece of work.
   - Use written language to explain issues, since verbal skills may be less apparent.

- Use exhibitions to obtain local market intelligence and feedback.
2. Ensure documentation is culturally adapted and not literally translated:
   - Brochures should be fully translated into local language on advice from local contacts.
   - Publish new catalogues in the local culture.
   - Set up the web site in the local language, with the web manager able to respond to any leads – after all, if a prospective client is expecting a fast response, waiting for a translator to arrive is not the way to provide it.
   - Adapt the titles of the services offered to match local perceptions.
   - Emphasise added value services.
3. Depending on the country or countries being targeted, operate as, for example, a European or Asian company, rather than a British company with a multilingual approach – think global, act local:
   - Arrange a comprehensive, multi-level programme of visits to the country.
   - Set up a local subsidiary company or local office or, failing that, set up a foreign desk inside the head office operating as if it is in the foreign country (keeping foreign hours, speaking foreign language etc.).
   - Change the culture of the whole company at all levels from British to European, Asian etc. as relevant.
   - Recruit local agents that have been educated in the UK, so they have a good understanding of UK culture too.
4. Implement a whole company development strategy:
   - Language strategy should be an integral part of a company's overall strategy as a learning organisation.
   - Identify the few individuals who can learn languages quickly and build on this.
   - Create in-house language provision.
   - Set up short-term student placements in the UK for foreign students, via a sponsored scheme such as the EU Leonardo da Vinci programmes.
   - Target markets where specialist language ability gives a competitive edge, e.g. China.
5. Subcontract the whole export process to a specialist company:
   - Hire a company to provide an export package of contacts, liaison, translation, language training, etc.

6. Pool resources with other companies:
   - Share language expertise and expenses with other companies.
7. In joint ventures, collaboration can be based on:
   - Equity/operating joint ventures, in which a new entity is created to carry out a specific activity. Seen as a long-term commitment, the new entity has separate legal standing.
   - Contractual ventures, in which no separate entity is created and instead firms co-operate and share the risk and rewards in clearly specified and predetermined ways. On the face of it, this form of joint venture appears to be more formal.
8. Management contracts:
   - The transfer of managerial skills and expertise in the operation of a business in return for remuneration.

Conclusion

With the advent of electronic communications, the possibilities that exist for quantity surveyors to operate on a European or global level have never been greater or easier to access. However, despite what some multinational organisations would have us believe, the world is not a bland homogeneous mass and organisations still need to pay attention to the basics of how to conduct interpersonal relationships if they are to succeed.

References

Cecchini, P. (1988). *The European Challenge*. Wildwood House.
Hofstede, G. (1984). *Culture Consequences: International Differences in Work-Related Values*. Sage Publications.
Hofstede, G. and Bond, M. H. (1988). *The Confucius Connection: From Cultural Roots to Economic Growth*. Organizational Dynamics.

Further reading

Bardouil, S. (2001). Surveying takes on Europe. *Chartered Surveyor Monthly*, Jul/Aug, pp. 20–22.
Brooke, M. Z. (1996). *International Management*, 3rd edn. Stanley Thornes.

Button, R. and Mills, R. (2000). Public sector procurement: the Harmon case. *Chartered Surveyor Monthly*, May, 24.

Cartlidge, D. (1997). It's time to tackle cheating in EU public procurement. *Chartered Surveyor Monthly*, Nov/Dec, pp. 44–45.

Cartlidge, D. and Gray, C. (1996). *Cross Border Tendering for Public Sector Building Work within the EU*. European Procurement Group, Robert Gordon University.

Commission of the European Communities (1985). *Completing the Internal Market*. Office for Official Publications of the European Communities.

Commission of the European Communities (1994). *Public Procurement in Europe: The Directives*. Office for Official Publications of the European Communities.

Commission of the European Communities 1998. *Single Market News – The Newsletter of the Internal Market DG*. Office for Official Publications of the European Communities.

Commission of the European Communities (2001). *Interpretative Communication*. Office for Official Publications of the European Communities.

Davison, L. *et al.* (1995). *The European Competitive Environment*. Butterworth-Heinemann.

DGXV (2000). *Analysis of the Electronic Public Procurement Pilot Projects in the European Union*. PLS-Ramboll.

Hagen, S. (ed.) (1997). *Successful Cross-Cultural Communication Strategies in European Business*. Elucidate.

Hagen. S. (1998) *Business Communication Across Borders: A Study of Language Use and Practice in European Companies*. The Centre for Information on Language Teaching and Research.

Hall, E. T. (1990). *Understanding Cultural Differences*. Intercultural Press.

Hall, M. A. and Jaggar, D. M. (1997). Should construction enterprises, working internationally, take account of differences in culture? *Proceedings of the Thirteenth Annual ARCOM 97 Conference, King's College Cambridge, 15–17 September*.

McKendrick, P. (1998). RICS Annual Report and Accounts: President's Statement. *Chartered Surveyor Monthly*, Feb, 2.

Tissier, M. *et al.* (1996). Chartered surveyors: an international future. *Chartered Surveyor Monthly*, Oct, 15.

Web sites

http://simap.eu.int/
www.ted.eur-op.eu.int
www.tendersdirect.co.uk

8

Education, research and practice

Marion Temple MA MA

Introduction

The preface to this book contains extracts from a number of reports that state that the industry perception of the quantity surveyor is of a person who:

- Fails to challenge established thinking
- Adds nothing to the construction process
- Lacks interpersonal skills
- Is unco-operative.

This chapter therefore looks at the role played by current approaches to education and research in relation to practice. In light of these perceptions, consideration is given as to how the current situation could be improved, and why it is important to surveying and to surveyors that it is improved.

Education and practice

From the surveying perspective, the past 30 years have seen the profession change from one where the principal route to qualification was part-time study to one for which admission is principally through study of an accredited university degree course. However, the potential beneficial impact of this largely graduate entry upon surveying and its professional practice has not yet been fully exploited. For too many current graduates, the moment that they bid farewell to their university upon graduation is the moment that they also bid farewell to meaningful links with their continued professional education.

The Royal Institution of Chartered Surveyors has a requirement that members must undergo a minimum number of continuing professional development (CPD) training hours each year. For too many, this is currently seen as a mechanistic process to be followed in order to place an appropriate tick in an appropriate box. One aim of this chapter is therefore to reinforce the benefits of CPD as a genuine continuation of education and learning undertaken by the professional so as to underpin and enhance their professional practice.

The Latham Report concluded that one of the root causes of the problems with UK construction was the silo approach to educating the various professional disciplines. From an early stage, architects are taught to cherish their *beaux arts* background and largely to ignore commercial and economic pressures, while quantity surveyors are taught to be suspicious of ... almost everybody else in the supply chain!

Previous chapters have dealt with the influence of cultural differences on good team working; similarly, undue specialism during the education and training of property professionals leads to cultural divides within the UK construction industry. A cultural adjustment is therefore essential if surveyors are to make a meaningful contribution to the interdisciplinary challenges facing the property industry, including that of sustainability in all its aspects.

Research and practice

One of the many pieces of damning statistical information to emerge from the current drive for a more professional approach to property and construction is that the UK construction industry's expenditure upon litigation exceeds the amount spent upon research and development. The vital importance of research to practice at the beginning of the twenty-first century is explored in this chapter. Examples of the benefits of research to practice are illustrated. In today's fast-moving world, the property professions must be willing and able to look to research in order to innovate and keep up to date.

Much property research today is conducted by private sector organisations, and the results will be exploited commercially. Traditionally this research has been viewed as private and secret, with little dissemination for the greater benefit of the profession or the industry. It is argued that this traditional approach is potentially positively harmful to property organisations in today's competitive business environment. The potential desirability of the organisation's losing some control over information in today's competitive economy is explained later in this chapter.

It is argued that research must be seen as, and must become, more inclusive rather than exclusive. This is one of the changes in approach necessary in order to make research a more immediate part of professional practice rather than a remote activity confined to an ivory tower. The traditional view of research, rooted in a post-medieval concept of the university as exclusively a community of scholars testing existing knowledge and discovering new knowledge purely for its own sake, must be challenged. Too often the silo approach demonstrated in property industry practice is replicated in education, where teaching and research are separated, sometimes even physically, as if to emphasise the view that academics are divided into 'teachers' and 'researchers'.

Research and practice in the knowledge economy

It is increasingly argued in the economic and management literature that successful organisations are those that seek to maximise the value of their intellectual property, and not those that seek to protect it most effectively. One interesting corollary of this approach is that information should be priced in terms of its value to the end user rather than in relation to the cost of its production to the provider. High-value information is that most needed and most valued by users, and is therefore the most profitable because of these characteristics.

This approach has been broadened by economists beyond its applicability solely to intellectual property in order to embrace other aspects of innovation, technology, competitive advantage and company and industry growth. For example, Shapiro and Varian (1999) argue that organisations should aim to maximise the value of their technology rather than to protect their technology. This is because, in today's dynamic world, the current technology itself is ephemeral. The organisation's competitive advantage derives primarily from the value derived from that technology through its exploitation in the marketplace. Shapiro and Varian (1999) put forward the following simple equation in order to explain their underlying hypothesis from the standpoint of the individual organisation:

Your reward = total value added to industry × your share of industry value

This means that the organisation can choose between two strategies: it can be protective of its knowledge and technology and gain a large share of a small market, or it can be open and gain a smaller share of a larger market. How should it choose? Where positive externalities exist, one advance in an industry will trigger positive feedback, leading to further positive externalities. This creates a virtuous circle of expansion in the industry and its market. So a strategy of being open can help to expand the industry and hence enhance the individual company's reward because its value expands along with the value added to the industry of which it is a component part. Increasingly, the evidence indicates that information and intellectual property are associated with positive externalities and positive feedback, leading to industry growth.

This model is relevant to many of the producer service industries that act as the current engines of economic growth in the world's richer economies. Where their customers are commercial or public sector clients rather than private individuals, large parts of the present-day property industry, including surveying, form part of this producer service sector. It is not clear that the growth potential to be derived from being part of this growth sector is currently being achieved by the surveying industry and its constituents.

In the opposite case, where organisations are protective and seek to control information, they can retain a large share of industry value, but are less likely to see the industry grow. It is this scenario that appears to better reflect much of the property and surveying industries today.

It is illuminating that the university sector, too, especially where it is successful in transferring new knowledge to industry, is primarily value driven rather than cost driven. For example, the study of US university practice reported in CVCP (1999) notes that technology transfer does not primarily provide universities with an additional income stream. Instead, the primary benefit is through the transfer of knowledge to the wider community in the public interest. The economic benefits are then reaped by the wider economy through the positive multiplier effects derived from the adoption of innovation. There is therefore little or no reduction in the direct cost of funding universities, but enhanced value results because there is an increased benefit to society through economic growth.

One hypothesis of this chapter is that surveying organisations have been unduly constrained by the culture of the traditional technology and, in seeking to control and protect information, have (probably unwittingly) held back the wider industry expansion from which they could all benefit. A more open approach to research and to connecting research to practice would benefit both research and practice. It can also be argued that a culture change in surveying is actually necessary to the industry's competitive survival in an increasingly global market.

Teaching, research and practice

This and the ensuing sections address the key questions raised by the discussion so far:

1. How does the model explained above impact upon university education?
2. How do the universities view the connections between research, teaching, students' learning and graduates' professional practice?
3. How does this relate to property and surveying education in the universities?

Despite many institutions asserting that the quality of their teaching is derived from the quality of their research, very few have a learning and teaching strategy that attempts to back this up ... in fact, most describe ways of developing teaching that has no connection with research. In contrast, Oxford Brookes University provides an example of a learning and teaching strategy that involves explicit attempts to strengthen the benefits, to both students and teachers, of being in an environment that supports research and consultancy (HEFCE, 2001).

Oxford Brookes University is currently home to research designed to investigate more fully these connections between a university's learning and teaching strategy and the provision of an environment that nurtures the connections between research, consultancy, teaching and practice. The author is associated with the LINK project that is exploring the *what, where* and *how* of linking teaching in higher education with research and consultancy, specifically within and across the planning, land and property management, and building-related disciplines. For the purposes of the present discussion, the emphasis will be upon the research project's outcomes in relation to land and property management. The wider context of the related disciplines will be touched upon because it is important in the context of the interdisciplinary approach to education, research and practice needed to complement the silo approach of single discipline education. The findings from the LINK project aim to enhance teaching–research–professional practice links to the mutual benefit of pedagogy, research, professional practice and the professional institutions, including surveying and the RICS.

Education: research and teaching strategies

This section considers some of the recent and current pedagogic research that has asked whether research should be linked to teaching for:

- All or some students
- At all or some levels of study
- In all or in some higher education institutions
- In all or in some subjects of study.

In summary, should the links between research, teaching, consultancy and professional practice be selective or be inclusive?

In favour of selectivity

Quantitative research conducted primarily in the USA during the early 1990s found little statistical correlation between 'good teachers' and 'good researchers'. In terms of content and methodology, the research also identified few connections between pure research and undergraduate teaching. Research was seen as primarily, if not exclusively, the domain of postgraduate research students. The results of this research would suggest that there are merits in a selective approach that categorises certain institutions and their staff as research-oriented, in contrast to other institutions and their staff that are mainly teaching-oriented. This division has been, and often continues to be, mirrored in policy and funding, which concentrates research resources into a limited number of identified institutions.

The extension of higher education in recent years to accommodate larger and more diverse student intakes makes it less likely that a research-driven curriculum is either desirable or feasible for undergraduate programmes of study. Larger class sizes and increased student numbers are associated with a greater distance between undergraduate study and specialist pure research. The selective model focuses upon quantitative research into the relationship between research and teaching and, interestingly, emphasises the quantitative, measurable outputs of that research, such as academic publications. This positive, quantitative approach is reflected in activities such as the current research assessment exercise in the UK.

The selective approach may therefore be categorised as output- and pure (as distinct from applied) research-oriented. A primary objective of research and of researchers within this framework is the advancement of pure knowledge and the ensuing product of that research, benchmarked through publication directed primarily at an academic readership. The selective approach is product, as distinct from process, oriented.

In favour of inclusivity

More recently, research has utilised a more qualitative methodology that indicates a mutually beneficial linkage between staff research, staff teaching and student learning. In the more recent US literature it is argued that research activity can and does serve as an important mode of teaching and a valuable means of learning. This emphasis upon research activity as distinct from research output indicates a broader, more process-oriented approach to research than that described above. Research-related skills, as well as research output, and their applicability to professional practice suggest a broader approach to the definition and interpretation of research within the context of the strategic development of contemporary higher education learning and teaching. The emphasis here is upon the normative attributes of research within the academic's role, in contrast to the positivist approach outlined in the previous paragraphs.

It has been found that students value research because research-active academics display current knowledge of and enthusiasm for their subject. This was also seen to add credibility both to the department and to the student's degree (Jenkins *et al.*, 1998). Further research at Oxford Brookes University, focusing upon the experience of taught postgraduate students, has endorsed the importance of the synergies between research and teaching. One important aspect of this more recent research is the implication for higher education of the growing number of taught postgraduate students who are either using postgraduate study to further their professional career development (outside an academic career) or are self-funding – or indeed both (Lindsay *et al.*, 2002). These students value staff whose research is perceived as being salient to the students' course content and hence the students' personal professional development – in other words, not necessarily pure research.

Research processes can actively inform student-centred and problem-based learning. The pedagogic links between research and teaching then become much richer than those of the student passively sitting in the lecture in order to hear the academic research expert's views and knowledge.

Selective and/or inclusive?

If the pedagogic focus is upon the student's learning experience, then research skills and processes can be integrated into learning from an early undergraduate level. One benefit of research-based learning is that it alerts all students to the nature of knowledge in their discipline and its developmental and contestable aspects. As noted above, higher education today is different from in the past in that there is now a wider range of subjects covered and an increased size and diversity in the student population. We have also noted that one feature of the society within which higher education takes place is now the rapid pace of the advancement of knowledge, its dissemination and application. The processes surrounding knowledge and its management are as important today as the knowledge itself. As a result, a broader-based approach to the definition of research can help illuminate the contribution of higher education to the non-traditional subject disciplines, of which surveying is one.

The skills required by graduates for life and employment in the new century are different from those needed a century ago. Today the emphasis is on the knowledge economy. This means educating students to have the creative skills that enable them to act independently, and to generate, apply and synthesise new ideas and knowledge throughout their working life. The contemporary body of knowledge is larger, more complex and more dynamic than ever, requiring different expertise and skills in order to use and manipulate it effectively. For this reason, Healey (2000) has argued that research skills for the academic should be both required and inclusive, to ensure that staff are at the forefront of both their subject and how to communicate it to students.

The inclusive model focuses upon the student's learning and the ways in which research-active and consultancy-active staff can enable students to acquire knowledge-related skills rather than solely to acquire knowledge.

Because today's students are embarking upon a path of lifelong learning through their continuing professional development, undergraduate (and much taught postgraduate) education has ceased to be the final point of education for many adults. Instead students need to be prepared for a life of learning, and this recognition has imbued higher education with learning outcomes that seek to ensure that its graduates are equipped as independent learners and self-developers. This in itself raises the profile of research skills and processes within higher education – for all students rather than for a few, and for all subjects rather than for only a few.

Education for surveying

How does the above argument, that university education strategies with respect to learning and teaching should positively enhance the links between research, teaching and professional practice, affect surveying education?

The existing literature base relating to the links between research and consultancy and teaching is primarily American. In relation to the property-related disciplines, these links have been discussed to some extent in the architecture and planning literature. However, there is little coverage in relation to surveying.

Because surveying education encompasses a wide range of subject methodologies, there is likely to be a wide range of teaching–research approaches and links even within cognate courses. For example, consider the wide difference in philosophy and methodology between valuation and law. To date, however, little research has been undertaken with regard to the relationships between research and teaching specifically in surveying education.

Of the subject disciplines covered by the LINK project, land and property management stands out for the absence of literature and discussion of its pedagogy and educational philosophy. As a result, we have little concrete evidence with respect to either the nature of the connections between research and teaching or the implications for the profession and professional practice.

Linking research to teaching

This link is most direct where academic staff research interests directly inform the curriculum content, with curriculum topics being drawn from current research as distinct from alternative criteria. The research ethos underlying the curriculum should be stronger within postgraduate than undergraduate programmes. However, for undergraduate programmes too, students need to develop and benefit from research skills. This is frequently reflected in the requirement for a dissertation or extended essay within the summative modules of undergraduate programmes, including those in surveying.

In addition to this example, opportunities to develop research skills in university education may be diverse. Examples range from a review of a selected part of the salient literature to devising a project outline or a response to a project brief, to analysing real data, using peer review within seminars, and presenting research results. Opportunities to develop and practise peer review skills are linked to vital professional skills of judgement and communication. It is therefore interesting that at least one professional body requires students to 'participate in a forum where their own view(s) are subjected to peer group criticism' as part of their learning experience (Construction Industry Board, 1998).

Linking consultancy and practice to teaching

At both undergraduate and postgraduate level, the student experience is enriched by links to consultancy and practice in order to add realism to the theoretical aspects of their education and enable students to understand the practical application of theory – and its limitations!

Students can attain a better understanding of the role of the consultant and of the links between their programme of study and professional practice where active professional experts are brought into the learning environment. Such professionals can discuss their professional experience with students through the mediation of the academic teacher. In addition, work experience and work-based learning contribute to the students' understanding of the realities and limitations of applying theory to practice.

Research-based learning

Interdisciplinary links

A key finding from multidisciplinary focus group discussions held under the auspices of the LINK project during 2000–2001 was that participants identified and valued the differences between the different disciplines and their cultures. For example, the professional culture and expectations of surveying/real estate management are clearly distinct from those of building/construction management.

Consequently, some of the identified links between teaching, research and consultancy and professional practice are discipline-specific. In this respect, some of the links identified through the project strengthen multidisciplinarity within which divisions between different bodies of knowledge and professional skills are retained and valued. Such a clear individual subject identity is plainly welcome as a counterpart to the strong identifiable professional culture embodied in the role of a professional institution such as the RICS.

In contrast, some identified examples of good practice and links are transferable across individual subject areas and so contribute to an interdisciplinary and potentially interprofessional approach to the study of property. A different perspective upon study material is offered by the use of examples linked to underlying issues such as the environment, development or sustainability.

The LINK project has yielded a rich matrix of practice, the discipline-specific being interwoven with the interdisciplinary. This indicates a real strength for the property professions and their development once a balance between an individual culture, such as that of surveying, and an understanding of the wider interdisciplinary property context can be attained.

Implications for stakeholders

We have argued that today's knowledge-based economy requires research-based learning to be extended to all students and staff in order for higher education to make an effective contribution towards fulfilling the needs of society. Only through a universal approach can research-based learning be widely disseminated and research skills embedded in graduates.

The LINK project has explored ways in which this ambition may realistically be achieved in today's universities. For many universities in the UK, developing links between research and teaching is important because they are also strengthening their research bases and expanding both taught and research postgraduate activity as a complement to existing strong undergraduate teaching. Also, for many universities the nature of postgraduate work is shifting in such a way that an increasing proportion of postgraduate teaching and research is connected to students' attainment of professional qualification and experiential development.

So what are the implications for the main stakeholders including employers and the professional institutions?

- *Students*: Students should be encouraged to understand, appreciate and utilise the links between research, practice and teaching. As learning participants and as users of teaching materials, students can be deployed in research tasks and projects as well as in activities related to professional practice. These learning activities can help students better to understand the role of research, the use and application of research skills within their professional employment, how the frontiers of knowledge expand in their subject discipline, and how to manage that knowledge.

- *Employers*: Because surveying courses are primarily vocational, the needs of employers and the employability of graduates impact upon the curriculum and its delivery. Where employers positively prefer graduates to have knowledge of professional practice, of current research and of research skills, then university education that fosters these attributes is more likely to emerge.
- *Professional institutions*: The external environment can also affect individual university courses through the influence of professional bodies such as the RICS and the guidelines issued by them with respect to curriculum content and student competencies. This influence is currently exercised by the RICS through partnership arrangements with its partner universities.

Enhancing surveying practice

The wide variety of examples pertinent to surveying and property is aptly illustrated by the case studies on the Association of Universities Research and Industry Links (AURIL) web site, where relevant cases range from environmental assessment to pedestrian flows and the design of safe landfill sites.

The business applications of property-based research are sometimes useful to the property industry and also to private industry and the public sector more generally, both nationally and internationally. For example, project management techniques developed initially for the construction industry can have wider applicability. Risk-analysis methodologies developed by academics in the early 1990s are embedded within decision-making procedures in tendering in both commerce and where public expenditure is involved (see, for example Raftery, 1994).

A further, more detailed, example is included here in order to demonstrate how research can help property professionals. *The Green Guide to Specification* (Anderson *et al.*, 2001) offers a tool that enables architects, surveyors, building managers and property owners seeking to reduce the environmental impacts of building materials to make informed choices. This is achieved by providing clear information as to whether the materials and components that are being specified for a building have a low or high environmental impact. A simple ABC rating system enables property professionals to choose the lowest environmental impact materials for construction and refurbishment projects from a wide range of specification options provided for all the principal building elements, including external and internal walls, floor systems, floor finishes, roofs, windows, doors, ceilings, paints, insulation, and landscaping. *The Green Guide*'s environmental profiling system therefore offers guidance for designers, property owners and facilities managers on the relative environmental impact of over 250 building materials and components.

This environmental profiling system was designed by staff in the Department of Real Estate Management at Oxford Brookes University in collaboration with a profit-making organisation and in partnership with the Building Research Establishment. Whilst full lifecycle assessments are complex, time-consuming and expensive processes, the environmental ratings in *The Green Guide* provide a simple but reliable way for designers and specifiers to assess their options, based on carefully researched, quantitative data.

As a part of the Building Research Establishment Environmental Assessment Method (BREEAM) programme, *The Green Guide to Specification* is a highly pertinent example here because it demonstrates the benefits of collaboration between business, the property professions, and academic and research institutions.

The system was designed to be used by busy professionals involved in the day-to-day management of building projects. For this reason, it had to be both easy to use and based on reliable quantitative data: essential criteria for the system to be accepted as credible by practising professionals and their clients.

The university environment offered the necessary academic rigour required for such a project. The commercial property organisation involved was able to contribute experience in project management, and an understanding of how property professionals best access and use information and guidance. For this project, the BRE was able to contribute a different research perspective from that of the university, further enhancing and widening the research and professional base on which the project was built.

The resulting project focused upon the key interdisciplinary issue of sustainability in construction, yielding potential benefits to a number of professions and authored by a cross-disciplinary team. It is therefore a pertinent example of the value to be derived from interdisciplinary research and practice outside the individual silo of the individual profession.

In addition to commercial applications, property-based research can also inform public policy across a range of topics, from health and safety specifications to the valuation, legal and pricing implications of land contamination, and the application of price mechanisms to the property markets in the emerging economies and in the transition economies of central and eastern Europe.

Knowledge flows

This section draws together the earlier discussions by considering the roles of the universities and industry in generating knowledge and encouraging its application in the property industry. An important component of knowledge generation within an economy takes place within the universities. If the economy in general and, in the context of the current chapter, surveying in particular is to benefit from this generated knowledge, then we need to understand how it flows from the universities into society and the property industry. Once we understand these flows, they can be improved both in relation to the rate of flow of new knowledge into surveying and in relation to the size of that knowledge flow.

In practice, five components of the flow of knowledge from the universities into the industry and the community can be identified:

1. The impact of academic research upon undergraduate teaching – a flow that is significant for its width as distinct from its depth
2. The academic research brought into industry by postgraduates – a smaller but deeper flow
3. The research derived from partnership between universities and industry
4. The knowledge derived from conferences and publications
5. The knowledge derived from pure research in the universities – the narrowest of the flows.

In general, most of the relevant research undertaken has focused upon the third of these flows: university–industry relations (UIRs) through research collaboration. This collaboration is seen as vital to the effective capitalisation of knowledge, which is increasingly seen as a prerequisite to economic growth (Blankenburg, 1998).

In spite of the increasing profile of graduate and postgraduate entry into the surveying profession, it is interesting that relatively little research has been undertaken into the size of the first two knowledge flows listed above. Nor has there yet been extensive explicit research into the surveying employers' view of the usefulness of the knowledge embedded in graduate recruits – although this can be deduced to some extent from employment patterns. The knowledge flow from the universities into surveying will only become optimal when it is better understood, and the RICS among others has a potentially important facilitative role to play in this respect.

The benefits of an interchange of people between the universities and industry have been rehearsed elsewhere. If we accept that a significant part of this interchange comprises the flow of graduates and postgraduates from the universities into surveying, then the flow of people will be essentially unequal, with a greater flow from the universities into surveying than in the opposite direction. So we can envisage the following scenario:

- The surveying industry resources and sponsors research and therefore the generation of additional new knowledge in the universities
- There is a positive impact upon the universities' teaching of their students
- Students graduate from the universities with enhanced knowledge, which they communicate to their employers in surveying
- The industry benefits from this enhanced knowledge and so is able to fund further research.

This generates a virtuous circle of economic growth for the surveying industry generated by the positive externalities associated with the knowledge economy, which are increasingly identified in the economics of innovation but are not yet often applied within surveying.

Conclusions

Little current evidence is available to inform us where surveying stands with respect to learning and teaching strategies for the delivery of the discipline within contemporary higher education. It is hoped that this chapter will help to inform and stimulate debate and discussion in order to remedy this deficiency and enhance both academic and professional practice.

It has been argued that, for the knowledge-based society of the new century, research-based learning should be universal in order for society to benefit from the enriched human capital associated with more, and more effective, education. In this context, it is proposed that a universal and inclusive, rather than solely a selective, model of research is especially pertinent. Furthermore, it is proposed that the synergies to be obtained from exploiting the links between research, consultancy, professional practice and teaching are an important component of enriching the student experience in higher education. These synergies provide a vehicle for helping academic teachers to help students to learn in an increasingly complex and dynamic environment.

There are clear benefits to surveying if it can help to enable its own expansion through capturing the advantages of a greater knowledge flow between the education sector and the industry. This chapter has argued that there are a number of mechanisms through which this objective could be achieved, ranging from greater research collaboration between the universities and the surveying industry to more active encouragement of knowledge acquisition and knowledge management skills by students, including those undertaking continuing professional development.

Acknowledgements

The LINK project is funded by the Higher Education Funding Council for England and the Department for Higher and Further Education, Training and Employment under the Fund for the Development of Teaching and Learning Phase 3. The project is being conducted by Professor Roger Zetter (Project Director, School of Planning, Oxford Brookes University), Dr Bridget Durning (Project Manager, School of Planning, Oxford

Brookes University), Professor Alan Jenkins (Project Advisor, Oxford Centre for Staff Learning and Development, Oxford Brookes University), Nick Bailey (School of the Built Environment, University of Westminster), Ron Griffiths (Faculty of the Built Environment, University of the West of England), Marion Temple (School of Architecture, Oxford Brookes University) and Pat Turrell (School of Environment and Development, Sheffield Hallam University).

The author is grateful to members of the Project Management Group and Project Steering Group for contributions to project meetings and informal discussions that have enriched this book. Thanks are due to Alan Jenkins, Roger Lindsay and their co-authors for permission to use as yet unpublished material. In addition, the positive contribution to the project outcomes made by the many contributors to the website and participants in the focus group discussions at the partner universities is gratefully acknowledged. Any errors or omissions are the sole responsibility of the current author.

References

Anderson, J., Shiers, D. and Sinclair, M. (2001). *The Green Guide to Specification*, 3rd edn. Blackwell Science.

Blankenburg, S. (1998). *University–Industry Relations, Innovation and Power*. Centre for Business Research, University of Cambridge Working Paper no. 102.

Committee of Vice Chancellors and Principals of the Universities of the United Kingdom (CVCP) (1999). *Technology Transfer: the US Experience*.

Construction Industry Board (1998). *Degree Courses in Construction and the Built Environment: Common Learning Outcomes*.

Healey, M. (2000). Developing the scholarship of teaching in higher education: a discipline-based approach. *Higher Education Research and Development*, 19(2), 169–89.

HEFCE (2001). *Strategies for Learning and Teaching in Higher Education*. Report ref: 01/37.

Jenkins, A., Blackman, T., Lindsay, R.O. and Paton-Saltzberg, R. (1998). *Teaching and research: student perspectives and policy implications*. Studies in Higher Education, 23(2), 127–41.

Lindsay, R., Breen, R., and Jenkins, A. (2002). Academic research and teaching quality – the views of undergraduate and postgraduate students. *Studies in Higher Education*, 27(3).

Raftery, J. (1994). *Risk Analysis in Project Management*. E and FN Spon (Chapman and Hall).

Shapiro, C. and Varian, H. R. (1999). *Information Rules*. Harvard Business School Press.

Web sites

www.brookes.ac.uk/LINK – FDTL project into the LINK between teaching and research and consultancy in planning, land and property management, and building.

www.york.ac.uk/org/auril – Association for University Research and Industry Links

Appendix

The following pages show examples of notices to be posted by public sector contracting authorities in order to comply with EU Public Procurement Directives (see Chapter 6).

These notices (source: TendersDirect) have been published in the *Official Journal*, and all relate to the same contract – i.e. the provision of project management and quantity surveying services for University College London. It should be noted that had the purchase authority been other than UK based, then the majority of the notices would have been published in the authority's native language.

Figure A.1 shows a Prior Information Notice (PIN), which is published either at the beginning of the year or, more usually in the case of construction contracts, when the project is first planned. A PIN notice is designed to provide contractors and professional services companies with advance information that a project is about to be put out to tender so that they can begin to prepare their response.

Figure A.2 shows an Invitation to Tender Notice, which provides information on the services required as well as contact details for the purchasing authority and the date by which potential suppliers must have responded.

Figure A.3 illustrates a Contract Award Notice, which is required after the contract has been awarded, and lists the name and address of the successful supplier.

Prior-information procedure

Title:	UK-London: project management and quantity surveying services
Purchase Authority:	UNIVERSITY COLLEGE LONDON
Journal Ref:	123386-1999
Published on:	02-Sep-1999
Deadline:	This notice expired on 19/08/00.
Contract Type:	This is a service contract.
Country:	United Kingdom
Notice Type:	Prior-information procedure
Regulations:	This document is regulated by the European Services Directive 92/50/EEC.

Tender Details

```
1. Awarding authority: University College London, Estates
and Facilities Division, Gower Street, UK-London WC1E 6BT.
Tel. (01 71) 391 12 41. Telex 28722 UCPHYS-G. Telegraphic
address:
University College London. Facsimile (01 71) 813 05 24.
2. Intended total procurement (Annex I A): CPV: 74142100,
74232400.
CPC reference No 867.
Project management and quantity surveying services.
University College London will be seeking expressions of
interest from consortia for the provision of the above
services for a contract to rebuild and refurbish a building
in Huntley Street, UK-London. The proposed building is to be
5 storeys, plus basement and sub-basement.
The value of the building contract will be in the order of
25 750 000 EUR (17 000 000 GBP) and service providers'
contracts related accordingly.
3.
4. Other information: The estimated date for the awarding of
these contracts will be 12/1999.
Additional information may be obtained from the address in
1.
5. Notice postmarked: 19. 8. 1999.
6. Notice received on: 23. 8. 1999.
7.
```

Figure A.1 Prior Information Notice.

Title: UK-London: project management and quantity surveying services

Purchase Authority: UNIVERSITY COLLEGE LONDON

Journal Ref: 127831-1999

Published on: 15-Sep-1999

Deadline: This notice expired on 15/10/99. Click here to view details of the award.

Contract Type: This is a service contract.

Country: United Kingdom

Notice Type: Invitation to Tender Notice - Restricted Procedure

Regulations: This document is regulated by the European Services Directive 92/50/EEC.

Tender Details

1. Awarding authority: University College London, Estates and Facilities Division, Gower Street, UK-London WC1E 6BT. Tel. (01 71) 391 12 41. Telex 28722 UCPHYS-G. Telegraphic address:
University College London. Facsimile (01 71) 813 05 24.

2. Category of service and description, CPC reference number, quantity, options: CPV: 74142100, 74232400. Category 12, CPC reference No 867.

Project management and quantity surveying services from consortia for a contract to part rebuild and part refurbish a building in Huntley Street, UK-London. This building was built in the early 1900s. It is proposed to part rebuild on the site and part refurbish the existing building to form research laboratories, associated offices, ancillary areas and some teaching areas. The building will be 5-storeys plus basement and sub-basement.

The value of the building contract will be in the order of 25 750 000 ECU (17 000 000 GBP). The service providers' commission will be from 2-3 years, and contracts will be related in value accordingly. It is envisaged that a maximum of 5 consortia will be shortlisted and invited to tender. The consortia should agree a lead consultant to respond to this notice.

3. Delivery to: UCL, UK-London.

4. a) Reserved for a particular profession: UCL require that organizations undertaking building consultancy work employ on that work staff having the relevant professional experience and competence for that job.

4. b) Law, regulation or administrative provision: The requirement in 4(a) is an administrative provision of the university.

4. c) Obligation to mention the names and qualification of personnel:
The names and professional experience of the staff to be responsible for the execution of the services will be sought from those invited to tender.

Figure A.2 Invitation to Tender Notice.

5. Division into lots: The service provider can only tender for the whole of the services. Where different disciplines form a consortium, the lead consultant should repond to the notice.

6. Number of service providers which will be invited to tender: A maximum of 5 consortia will be shortlisted and invited to tender.

7. Variants: Variants will not be accepted.

8. Time limits for completion or duration of the contract, for starting or providing the service: 2-3 years.

9. Legal form in case of group bidders: Joint and several liability.

10. a)

10. b) Deadline for receipt of applications: 15. 10. 1999.

10. c) Address: Mr T. Edwards, UCL, Estates and Facilities Division, 1-19 Torrington Place, UK-London WC1E 6BT.

10. d) Language(s): English.

11. Final date for the dispatch of invitations to tender: 30. 10. 1999:
this date may change subject to level of initial response.

12. Deposits and guarantees: No deposits or guarantees required.

13. Qualifications: Proof of the service provider's financial standing must be furnished by:
appropriate statements by bankers;
submission of latest audited annual report and balance sheet;
statement of turnover, and turnover in the previous 3 financial years;
a list of relevant principal contracts undertaken in the past 5 years,
details to include values, dates, description and clients;
description and evidence of quality-control procedures;
indication of membership of technical bodies of those staff responsible for the provision of the service;
details of 2 current clients, their names and addresses, who are willing to supply references;
evidence of professional-indemnity insurance.

14. Award criteria: Relevant experience of tenderers and professional competence, quality and technical merit of tenders. The economically most advantageous tender.

15. Other information: If invited, tender bids will be in sterling.

16.

17. Notice postmarked: 25. 8. 1999.

18. Notice received on: 6. 9. 1999.

19.

Figure A.2 Continued

Contract Award Notice

Title:	UK-London: project management and quantity surveying services
Purchase Authority:	UNIVERSITY COLLEGE LONDON
Journal Ref:	51239-2000
Published on:	21-Apr-2000
Deadline:	N/a.
Contract Type:	This is a service contract.
Country:	United Kingdom
Notice Type:	Contract Award Notice
Regulations:	This document is regulated by the European Services Directive 92/50/EEC.

Tender Details

```
1. Awarding authority: University College London, Estates &
Facilities Division, Gower Street, UK-London WC1E 6BT.
Tel.: (020) 76 79 12 41.
Telex: 28722 UCPHYS-G. Telegraph: University College London.
Fax: (020) 76 79 05 24.
2. Award procedure chosen, justification (Article 11(3)):
Restricted.
3. Category of service and description, CPC reference
number, quantity:
CPV: 74142100, 74232400.
Category 12; CPC reference No 867.
Project management and quantity surveying services.
4. Date of award: 12/2000.
5. Award criteria: Selection based on weighted scores for
the following:
competence, quality and technical merit of tenderer,
economically most advantageous tender.
6. Tenders received: 41.
7. Service provider(s): Mace.
8.
9.
10.
11.
12. Contract notice published on: 15.9.1999.
1999/S 179-127831.
13. Notice postmarked: 7.4.2000.
14. Notice received on: 11.4.2000.
15.
```

Figure A.3 Contract Award Notice.

Index